Vinzenz Kratzer
Governmental Migration Research in Germany

AF849720

Culture and Social Practice

*To Aga, Zosia, and Paula*

**Vinzenz Kratzer**, born in 1985, conducted his PhD research at the European University Viadrina in Frankfurt/Oder. As a holder of the Polish-German Double Degree in Political Science and European Studies, Vinzenz Kratzer focusses on political and ethnographic studies of migration in Central and Eastern Europe.

Vinzenz Kratzer

# Governmental Migration Research in Germany

Knowledge Production at the Federal Office for Migration and Refugees

[transcript]

This work has been accepted as PhD thesis by the Europa-Universität Viadrina Frankfurt/Oder.
Examiners: Werner Schiffauer, Michał Buchowski

Funding: Friedrich-Ebert Foundation (PhD Scholarship), Europa-Universität Viadrina

**Bibliographic information published by the Deutsche Nationalbibliothek**
The Deutsche Nationalbibliothek lists this publication in the Deutsche Nationalbibliografie; detailed bibliographic data are available in the Internet at http://dnb.d-nb.de

**© 2021 transcript Verlag, Bielefeld**

All rights reserved. No part of this book may be reprinted or reproduced or utilized in any form or by any electronic, mechanical, or other means, now known or hereafter invented, including photocopying and recording, or in any information storage or retrieval system, without permission in writing from the publisher.

Cover layout: Maria Arndt, Bielefeld
Proofread: Molly Wimmer
Printed by Majuskel Medienproduktion GmbH, Wetzlar
Print-ISBN 978-3-8376-5709-8
PDF-ISBN 978-3-8394-5709-2
https://doi.org/10.14361/9783839457092

Printed on permanent acid-free text paper.

# Contents

# What Makes Knowledge Governmental?

September 2015. It is a hot, sunny day at the end of the "refugee summer" which strangely combined the end-of-summer laziness with a state of particular emergency conveyed by the media. For weeks, the news was dominated by reports about a massive influx of refugees and the resulting break-down of the registry mechanism for new arrivals. The *Bundesamt für Migration und Flüchtlinge* (BAMF), or Federal Office for Migration and Refugees, constituted as the central authority responsible for registration and conduct of asylum processes the epicenter of bureaucratic activity and media criticism during that time. Though I had arranged interviews with BAMF officials for my research project some months ago, I anxiously reconfirmed the appointment a week in advance, almost expecting them to be cancelled due to the latest developments. To my relief, they were not.

The Federal Office for Migration and Refugees lies on an arterial road with heavy traffic south of the center of Nuremberg. It is a large, four-story building spanning over several hundred meters along the road. Originally, the building was erected in 1939 as a barrack for the SS in the immediate vicinity of the National Socialist Party rally area. Today, the building does not reveal much of this history; nevertheless, the uniform brick and granite facade radiates a stern, bureaucratic purpose.

I arrive by car at the BAMF and am greeted by a sign – "Entering Strictly Prohibited" – at the front gate. The security measures are tighter than I had anticipated: there is a security booth by the entrance of the building and a locked double-door entrance. Judging by the visual impression, both have been added relatively recently. After registering with my ID at the security checkpoint, I am given a visitor's badge which I must wear at all times.

The entrance area was artistically designed to make a reference to the history of the building, a graphic table on the wall presenting it. This artwork and the building's history seem to be a standard small talk item for visitors: both people picking me up at the entrance on two consecutive days made almost identical remarks about the enormity of the complex, how easy it is to get lost in the long hallways, and how the artistic design of the entry hall deals with the building's problematic past.

Subconsciously, I expected the state of emergency as conveyed by media to be visible in the physical center of the migration policy system in Germany: some trace of the towering mountains of asylum files waiting to be decided upon, a visual expression of bureaucratic chaos, or at least government officials with files tucked under their arms hastily moving between offices. However, at the site, there is no particular emergency or busy frenzy visible. While I wait to be picked up by my interview partner in the entrance hallway, the lunchtime traffic slowly begins, with small groups of officials chatting about this and that while walking towards the cafeteria. In sum, the scene looks like any other day in any random mid-level federal German government building.

(Field notes, September 2015)

## Knowledge Production and Migration Policy Making

The Federal Office for Migration and Refugees is the central executive authority for federal integration and migration policy measures in Germany. It assumed this role relatively recently, as a result of the fundamental migration policy and administrative reforms in 2005 often referred to as a "paradigm change."[1] This change was triggered by the then newly elected Red-Green Government, which was eager to introduce reform to a policy-field with a decade-long history of political stalemate. At the same time, a rising trend of "evidence-based policy-making" since around the turn of the millennium provided the general context to this political effort.[2] In 2001, the *Unabhängige Kommission Zuwanderung* (Independent Commission Immigration) was founded to formulate scientifically grounded reform proposals for legal and administrative aspects of migration policy.[3] The Independent Commission planned to turn migration into a policy field steered by expert knowledge,[4] following the intention to "move away from policy based on 'dogma' to 'sound evidence' of 'what works'."[5]

The Independent Commission's final report clearly stresses the merits of knowledge-informed policy-making:

> "The acknowledgement of reality has replaced political taboos. Increasingly, public debate is governed by rationality. Germany needs [...] both permanent and temporary migration for the labor market [...]. How many migrants should come is decided by the polity with the support of the Immigration Council."[6]

However, sharing the fate of many similar reform attempts, the Independent Commission's propositions were largely ignored in the subsequent legisla-

---

1 Engler 2014, p. 67, Kreienbrink and Worbs 2015, p. 330, Bade 2017, p. 198

2 Sanderson 2002

3 Schneider 2010, Scholten 2011b, 255f.

4 Expert knowledge in this context refers to knowledge arising from scientific knowledge production (Cp. Boswell 2009b, p. 4).Knowledge production, in turn, is usually used interchangeably with "expert knowledge" or "research" in the relevant literature (Cp. for example Scholten et al. 2015b, Boswell 2009b, p. 4). In this text, knowledge production will be used in this sense (Cp. Bourdieu 1977).

5 Boswell 2009b, p. 3

6 Unabhängige Kommission "Zuwanderung" 2001, p. 1

tive process.[7] In particular, the Residence Act was purged from most features which ensured the systematic inclusion of scientific expertise into policy-making. Instead of an independent research institute and the Immigration Council of experts mentioned in the quote above, an in-house Research Group was established at the Federal Office under tight administrative control.[8] The knowledge production of the BAMF's Research Group will be analyzed in this thesis.

This history is of some relevance for the analysis of governmental knowledge production, since both the Independent Commission's recommendations about the future role of knowledge in policy-making as well as the failure of implementing them represent two competing theory streams. In relation to the former, in government documents and among researches, an instrumental approach to knowledge production is prevalent according to which knowledge is a key resource for political action.[9] Knowledge in this sense is used as a source of information, as a means of enhancing output quality, or as a source of legitimization.[10] This concept can be traced back to Max Weber's idea of bureaucracy as rule through abstract, impersonal decisions which require technical knowledge on the side of the government official.[11] This understanding is mirrored in the legal text describing the tasks of the Research Group as "conducting scientific research on migration issues (accompanying research) with the aim of obtaining analytical conclusions for use in controlling immigration."[12] The instrumental approach of knowledge utilization relates closely to the concept of *Ressortforschung* (Departmental research), a government-sponsored branch of applied research.[13]

However, the ultimate failure of the Independent Commission's reform proposals can be connected to the observation that in practice, policy-making is rarely guided by the ideal of scientifically grounded decision-making. This is even true for knowledge that has been directly commissioned by the

---

7 Cp. Schneider 2010, 277ff.

8 Bade 2001, p. 32

9 This understanding follows Boswell 2009b, p. 5 in her use of the term instrumental knowledge which includes various approaches sharing the assumption that knowledge is valued primarily for its informational content.

10 Cp. Schneider 2010, 74ff.

11 Weber 2005, 185ff.

12 Quoted from §75(4) Residence Act

13 Cp. Barlösius 2008, Groß 2010, Lundgreen 1986

government, such as the Independent Commission's report.[14] The constant deviation between rhetorical praise of scientific knowledge and actual political practice gave rise to criticism of the mainstream instrumental approach.[15] Several critical accounts aim at explaining this gap: these include a steering-pessimistic system-theory approach, according to which communication errors between the two incompatible systems "politics" and "science" lie at the root of this phenomenon.[16] Similarly, sociology of science studies often point to a cultural difference between science and politics which impedes the correct implementation of scientific knowledge.[17] In her often quoted study on knowledge production in the BAMF, Christina Boswell explains the lack of influence of the BAMF's knowledge production with alternative uses of the knowledge in policy-making, most importantly ex-post legitimization or substantiating decisions already taken.[18]

The knowledge production at the BAMF is however hard to capture with these critical concepts as well: On the one hand, the principal criticism of a lack of systematic influence on political decision is valid, since there is actually little evidence for proper instrumental knowledge use. On the other hand, however, critical theories seem to be unable to sufficiently explain exactly why: For example, the above-mentioned systems-theory approach states that systematic differences inhibit proper communication between politics and science as a matter of principle. However, as a detailed discussion of the history of the establishment of the Research Group will demonstrate, the assumption of a systematic policy-science gap cannot be easily maintained.[19] The Research Group managed to establish itself at the center of government migration research with some success. This is visible in the fact that it was able to secure more and more resources over time and established a position within the state administration and to some extent within academia as well. Today, the BAMF Research Group is among the most active publishers of migration and integration research in Germany.[20]

---

14 Boswell 2015, p. 36

15 See Boswell 2009b for a detailed discussion of the various alternative approaches to knowledge use in political decision making.

16 Nassehi et al. 2009, p. 7, Kraler and Perchinig 2017, p. 85

17 Boswell 2009b, p. 9

18 Boswell 2009b, 5ff., Scholten et al. 2015a, p. 318

19 Cp. for example Kraler and Perchinig 2017, p. 75

20 Schimany and Schock 2012, Leibnitz-Institut für Sozialwissenschaften 2010, p. 26

Furthermore, if the actual content of this research is analyzed, the major part of the BAMF's studies cannot be attributed to neither symbolic nor instrumental uses in Boswell's understanding.[21] This is mainly because for the most part, clear policy recommendations are absent of the Research Group's publications. Instead, many publications entail general socio-demographic information on particular target groups or describe institutional or legal arrangements in migration administration. This goes somewhat against what might be expected from a research institution which considers "the preparation, monitoring and evaluation of policy measures or programs"[22] as its core responsibility. As a result, most authors agree that the political use of the major part of the BAMF's research projects is "unclear."[23]

However, both instrumental and alternative approaches stand at odds with the self-perception of the Research Group: Despite complaints (usually off-tape) that they do not have much political influence,[24] BAMF researchers maintain that they do produce politically relevant scientific knowledge. In this context, the Research Group draws on a specific understanding of applied research,[25] which is discoursively constructed against theoretical, academic research:

> "We conduct academic studies, only the research question is usually not theory-driven, and that is a difference to universities. Here, we focus on applied research. [There is a wide array of] policy-counseling institutes which likewise follow an academic approach, which are almost always managed by academically trained scientists, but which have a more diversified audience."[26]

This self-understanding cannot simply be ignored: It might be true that knowledge production according to instrumental principles does not work

21 Boswell 2009b, p. 182

22 Kreienbrink and Worbs 2015, p. 329

23 Boswell 2009b, 187 f.. Cp. also Scholten et al. 2015a, 318f.

24 Cp. also Boswell 2009b, p. 5

25 Cp. also Barlösius 2008, p. 23

26 "Es sind akademische Arbeiten, [...] nur ihre Fragestellung ist in der Regel nicht theoriegeleitet, und das unterscheidet sich von dem, was an Universitäten passiert. [...] Bei uns steht die angewandte Forschung im Vordergrund. [Es gibt eine große Bandbreite an] politikberatenden Instituten, [...] die [...] auch akademischen Anspruch haben, die natürlich [...] fast immer von akademisch ausgebildeten Leuten geleitet werden, aber die ein breiteres Publikum haben."(Interview with a BAMF researcher, 2015)

in practice, but this does not mean that the knowledge production at the BAMF can be wholly dismissed as mere rhetoric or propaganda.[27] In the approach adopted in this thesis, the failure of knowledge production does not constitute the conclusion but rather the point of departure for analysis: instead of pointing out what knowledge production fails to do politically, the question is what it actually does instead.[28] Specifically, the thesis focuses on practical aspects of knowledge production, the political effects of the knowledge and the resulting epistemic features of governmental knowledge.

## Seeing Like a State

To make this practice-oriented approach productive for the analysis it is worth reflecting on the production conditions of knowledge and the way it is connected to governance. James C. Scott's *Seeing like a State,* a study on several large-scale agricultural modernization and development schemes, provides an inspiring point of departure for this task:

> "The premodern state was, in many crucial respects, partially blind; it knew precious little about its subjects, their wealth, their landholdings and yields, their location, their very identity. It lacked anything like a detailed 'map' of its terrain and its people. It lacked, for the most part, a measure, a metric, that would allow it to translate what it knew into a common standard measure necessary for a synoptic view."[29]

While Scott describes processes of mapping forests, creating cadastral maps, or establishing standard units of measurement and their implications for the exercise of political power, corresponding programs and processes in governmental migration research are apparent: After years of political neglect of the field, standard statistical compendiums were created and updated every year.[30] Newly developed meta concepts were applied by replacing the outdated German-foreigner dichotomy in statistics with a new concept (Migrant Background[31]) thereby introducing a standard measure similar to Scott's under-

27 Cp. Ferguson 1994, 17f.

28 Cp. also Foucault 2014, p. 80

29 Scott 1998, 17 f.

30 Cp. Bundesamt für Migration und Flüchtlinge 2006

31 Cp. Salentin 2014. For a definition of the term, see Statistisches Bundesamt 2007, p. 6.

standing. In integration research, a similar strategy towards the standardization of statistical parameters which describe and evaluate the progress of integration is discernible.[32] From the body of literature of government migration research, among the most-discussed studies in this time is the work "Muslim Life in Germany" which most importantly established an official count of Muslims on a national level for the first time.[33] In this context, Scott's approach accurately describes research projects that are hard to capture in the instrumentalism paradigm: Most of this knowledge cannot be connected to a specific political issue, and the according studies hardly ever contain recommendations for policy-making.

Such knowledge has otherwise often been described as lacking political relevance[34] or has been attributed to merely symbolic uses of knowledge in the literature.[35] In contrast, Scott demonstrates the political usefulness of such knowledge. Scott's approach connects crucial points between the immediate conditions and methods of knowledge production on the one hand and a greater picture about political power and the establishment of statehood on the other hand.

In this sense, Scott can be regarded as a representative for a research tradition which focuses on the various interconnections between governance and knowledge production. One of the single most important contributions in this line of thought is Foucault's concept of governmentality, according to which the execution of political power relies increasingly on mechanisms of self-steering and technical, particular knowledge exactly of the kind Scott writes about.[36] This approach is particularly useful for the analysis of governmental knowledge production since it avoids two main analytic traps: Firstly, by focusing on the interconnections between governance and knowledge production, the inadequate separation of "science" and "politics" is abandoned in favor of a holistic perspective. This is especially important for the analysis of the BAMF, which sees itself as a "boundary organization"[37] since it combines tasks and roles from "both worlds".[38] Secondly, the narrow focus on problem-

32 Cp. Bil and Verweij 2012, Friedrich and Waibel 2012
33 Bundesamt für Migration und Flüchtlinge 2009c
34 Cp. Kraler and Perchinig 2017
35 Cp. Boswell 2009b, p. 182
36 Foucault 2014, p. 17, Rose 1991, p. 675
37 Scholten 2011b, 46f.
38 Kreienbrink and Worbs 2018, p. 243

solving (or the lack thereof) through knowledge in the process of political decision making is broadened to include long-term and indirect political effects of knowledge production.[39] Knowledge in this understanding is not a somewhat objective input into politics, but rather formed and co-produced in the course of government. This does not mean that this knowledge can simply be regarded as propaganda, as stated above – the knowledge has to fulfill quite rigid methodological and coherence criteria.[40] Not coincidentally, the BAMF's methods of knowledge production resembles academic knowledge production quite closely in this regard. However, the most important quality criterion is not theoretical coherence or novelty, but rather the question if the knowledge is useable for government or not. In this sense, the requirement of political relevance is the most important structural feature of governmental knowledge.

Following Scott, this study employs a practice-oriented concept of knowledge production. This perspective is inspired by the basic insight that knowledge does not simply emerge from objective facts, but rather has to be constructed and arranged in a particular way.[41] In this sense, knowledge production is neither "deliberate construction"[42] nor a straightforward discovery of ex-ante existing truths. This practice of knowledge production becomes apparent in the following quote of a government researcher describing their work:

> "We work flexibly with what serves best. [If] we have a concrete question, we look which methods we can use to answer the question posed to us. In this we are not overly committed to a specific theoretical concept. If we refer to definitions [e.g. in the National Migration Report], these relate to statistical data, and the statistical data depends on legal regulations."[43]

---

39 Ozga et al. 2009, 358f.

40 Rose 1991, Boswell et al. 2011

41 Amir-Moazami 2018b, p. 93. Cp. also Gibbons et al. 1994, p. 3

42 Schiffauer 2018

43 "Wir arbeiten flexibel mit dem was da ist. [...] Wir haben eine konkrete Frage, wir schauen uns an mit welchen Methoden wir die konkrete Frage die uns gestellt wird beantworten können. Und sind nicht übertrieben eng hinter [...] einem Theoriekonzept her. Wenn wir Definitionen benutzen [zB. im Migrationsbericht] richtet sich bei uns nach den statistischen Erhebungen, die statistischen Erhebungen wiederum richten sich nach dem was in unseren Gesetzen steht."(Interview with a BAMF researcher, 2015)

In sum, the production of politically relevant knowledge can be best understood like a pragmatic tinkering with the resources, institutional confines, and strategic opportunities at hand.[44] At the BAMF Research Group, this tinkering process is visible in many practical aspects of knowledge production: research questions are formulated in a negotiation process involving ministries, researchers, and the Federal Office's administration to balance out different interests of political relevance, scientific credibility, and resource efficiency. Data is usually used because it is available, not because it is especially valid.

This is above all true for the Research Group's main data source, the *Ausländerzentralregister* (Central Registry for Foreign Nationals, AZR). Data from this source is methodologically problematic because it excludes naturalized persons as well as a good share of EU-foreigners; as a consequence, the AZR contains a bias towards the "socio-economically least successful."[45] Despite this, AZR data is used extensively since it is exclusively available to the BAMF[46] and therefore constitutes a unique selling point for the BAMF's research. Another pragmatic aspect of knowledge production is the publication strategy which arose from a compromise between academic and bureaucratic practices. As a result, politically relevant knowledge is communicated strategically to maximize its political impact.[47] The most consistent publication strategy is the practice to gear publication towards avoiding negative feedback. This can be connected to the peculiar position of the BAMF which is frequently blamed for policy failures originating from higher hierarchical levels. To a degree, it also explains the above-mentioned restraint in policy recommendations.

Summing up, if knowledge *production* is read as a metaphor, two meanings of the term "production" become apparent:[48] On the one hand, production entails a connotation of synthetically, similar to Chomsky's notion of "manufacturing consent" with manipulative intent. On the other hand, production refers to industrial process organization, where multiple workers are organized to collaborate for the manufacturing of a given product. While the connotation of manipulation is surely not irrelevant, this thesis addresses above

---

44 Cp. also Latour and Woolgar 1986

45 Salentin 2014, p. 25

46 Kreienbrink and Worbs 2015, p. 327

47 Cp. Mayr et al. 2011

48 I follow here Boswell's similar discussion of the word "manufacturing". Cp. Boswell 2018

all the practical aspects of knowledge production in a bureaucratic system which resembles in a way the division of labor in a factory.

## Four Features of Governmental Knowledge

The BAMF Research Group has been introduced as a paradigmatic case highlighting the inconsistencies in the theory debate between "instrumentalist knowledge" and various "gap theories" explaining the lack of actual political influence of the knowledge generated. As an alternative, a practice-oriented research perspective was sketched out, focusing mainly on the interconnections between governance and knowledge production.

By itself, this focus on practical aspects of knowledge production is neither original nor surprising – after all, it follows broadly the insights from sociology of science studies according to which knowledge production is the construction of truth according to established methodological principles.[49] However, in the context of the state, this focus is particularly useful in highlighting the specific features of governmental knowledge as an outcome of the conditions of production conditions as well as policy aims. By and large, four basic characteristics of governmental knowledge can be distinguished:

First, governmental knowledge is politically relevant.[50] From the perspective of governmental researchers, political relevance is the key difference between governmental and academic forms of knowledge production. In the theoretical literature, however, this relevance is routinely denied on the grounds of according empiric evidence, as mentioned above. The practice-oriented approach reconciles both perspectives: Political relevance is conceptualized as a quality standard for knowledge production, similar to the requirement of theoretical coherence in academia. While it is hard to prove direct influence on single political decisions, political relevance has a decisive impact on the process and outcome of knowledge production. This is for example visible in long-term developments in the research focus: Initially, the Research Group largely drew up its own research agenda.[51] However,

49 Scholten 2011b, 29ff.

50 In this text, the terms "political relevance" and "practical relevance" are used interchangeably. This reflects the according use at the BAMF and the concept of political relevance discussed later in this chapter. Cp. also Harris 2015, p. 27

51 Boswell 2009b, p. 180

after the establishment phase, a strategy of systematically acquiring research mandates from state authorities is clearly discernible, therefore adjusting knowledge production to demand. Regarding the research perspective, a distinct policy relevance effect is visible as well which narrowed down from a broad, all-encompassing overview perspective to specific target groups which are subject to governmental intervention, particularly from the Ministry of the Interior. Furthermore, practical relevance is not a uniform standard of quality but rather a flexible requirement according to the actual practice for which knowledge is produced. In this thesis, four different aims of political relevance will be discussed: Administration, depoliticizing, calming public debate, and legitimization. Administrative knowledge is relevant in the sense of the above-quoted legibility concept, to introduce standard measurements for facilitating political steering ("what gets measured gets managed"[52]). Depoliticizing is an effect of framing originally political problems in technical terms.[53] At the BAMF, this effect is most visible in integration research: Initial research projects were targeted on broader societal issues such as "the impact of immigration on the German Society",[54] or the construction of an indicator system for integration.[55] These projects are meanwhile replaced by technical examinations of the impact of integration courses, or the examination of integration parameters of single legal status groups.[56] This narrow perspective contributes to the changed understanding of integration, which is framed not as a political problem for the society as a whole, but rather as a technical task for the appropriate government authorities, thereby depoliticizing the issue.

Another discursive function of knowledge is calming the public debate to silence overly racist, xenophobic or otherwise undesirable statements about migrants (Muslims in particular) by superior knowledge and "objec-

52 Karabell 2014, p. 13

53 Ferguson 1994, p. 256

54 Bundesamt für Migration und Flüchtlinge 2005a

55 The integration report series from 2008-2012, see also Bundesamt für Migration und Flüchtlinge 2009b, p. 31: "The aim of the integration report is [...] to display the immigration status of the immigration population in Germany to the broad public."

56 For example, immigrating spouses (Bundesamt für Migration und Flüchtlinge 2014b), Bundesamt für Migration und Flüchtlinge 2012i, university students (Bundesamt für Migration und Flüchtlinge 2012h) and graduates (Bundesamt für Migration und Flüchtlinge 2014a)

tive facts".[57] Finally, legitimization is a frequently described political use of knowledge if studies are conducted to bolster ex-ante a specific political claim or strategy.[58] Legitimizing knowledge is for example visible in the BAMF's dual concept of migration potential: In the understanding of the BAMF, migration potential entails both migrant action (potential *of* migrants) and structural migration pressure as an indicator for future movements (*potential* migration). However, these elements are employed context-specific, depending on regions of origin.[59] In this way, the dangers of *potential* African and Eastern European migration potential are underlined, whereas the economic potential *of* migrants from EU-countries is emphasized. In this way, the BAMF's understanding of migration potential mirrors neatly the EU's migration strategy of counterbalancing intra-EU freedom of movement with increased efforts to seal off the external borders.[60]

All in all, political relevance is the most important single feature to form and influence the knowledge production at the BAMF. It can thus be considered both the key difference to academic knowledge production and a valuable entry point for analysis if different potential uses for knowledge are considered.

Second, governmental knowledge is inert and structurally conservative. This structural feature follows both from the fact that knowledge is produced in a government authority and again from policy relevance considerations. Structurally, the most important reason for conservatism is according demand on the side of ministries: According to Barlösius, ministries prefer probed and uncontroversial knowledge over experimental, "peak of science" oriented research.[61] Arguably, this effect grows stronger with the degree of dissemination of a given publication.[62] This explains the numerous copy-pasted passages of the Migration Reports, the Research Group's flagship annual publication: These repetitions are not a corner-cutting strategy, but

57 Cp. Schepelern Johansen and Spielhaus 2018, p. 128, Boswell 2009b, p. 201
58 Cp. Boswell 2009a
59 Potential of Migrants: Bundesamt für Migration und Flüchtlinge 2014e; Migration Potential: Bundesamt für Migration und Flüchtlinge 2009d, Bundesamt für Migration und Flüchtlinge 2012a.
60 Bade 2013, p. 15, Kratzer 2018b
61 Barlösius 2008
62 Another effect causing structural conservatism are expected political controversies triggered by research on controversial topics such as the naturalization study discussed later in the text.

rather constitute the outcome of the multiple checks and editing levels this document passed in the course of its production.

Additionally, structural conservatism can be considered an outcome of the bureaucratic organization of knowledge production. Typical bureaucratic media of knowledge production such as memos, tables, statistics, and most importantly annual reports share a logic of accumulation: over time, additional data points are connected to a time series; the data thus becomes increasingly valuable.[63] At the same time, the introduction of a new concept becomes more and more costly since "you have to start with square one", as a government researcher explained.[64] Therefore, once appropriate indicators, categories and statistical concepts are established, they are remarkably stable. This can lead to effects similar to Beck's "zombie categories", when concepts stay rigidly in place even when the social phenomenon they allegedly describe have changed.[65] The "Guest Worker" framework illustrates this feature of governmental knowledge well: Based on a governmental report series on foreign labor recruitment from the 1960s, social research on foreigners quite consistently employed a five-country comparison scheme reflecting the numerically most important "Guest Worker" recruitment countries of origin (Turkey, Italy, Spain, Yugoslavia, Greece);[66] the data presented focused strongly on economic and related social features of the migrant population such as employment rate, income, household size, age and gender specifications, and so on.[67] In principle, the "Guest Worker" concept remained the leading category system until the concept of "Migrant Background" was introduced in 2005, thus creating the counterfactual impression of the foreign population in Germany as "Guest Workers" from former recruitment countries. This concept came under increasing pressure for the fact that only a declining share of migrants were actual workers, the fact that the share of the largest recruitment countries gradually declined, and the fact that the increasing share of naturalized foreigners rendered the legalistic German-Foreigner divide to a degree irrelevant. The effects of this conceptualization can be illustrated by the stubborn examination of foreigner's "propensity of return" almost forty years after the

63 Cp. Rose 1991, p. 673 for an overview over the role of numerical data in governance.

64 Research Notes, February 2017

65 Beck 2000, 16ff.

66 Mehrländer 1987, 89ff.

67 Bundesamt für Migration und Flüchtlinge 2010c, p. 25

last "Guest Workers" have been recruited; many of those alleged potential returnees have in fact never lived outside of Germany.[68]

Be that as it may, the "Guest Worker" framework of analysis illustrates quite well the impact of structural conservatism on knowledge production: It helps establishing a proper way of measuring and sorting. Different entities and population groups are quantified by establishing an order of essentially comparable units; furthermore, an agreement over the proper way to quantify and evaluate policies is established (most importantly the number of foreign workers and the unemployment rate).[69] Furthermore, the framing of knowledge in technical and at times boring and repetitive ways cannot be solely considered a deficit but rather a highly productive feature of governmental knowledge:

> "It takes hard discoursive work to keep things as they are. Making the world seem stable when it is in fact in constant flux means that wielding power involves the ability to freeze meaning. This has to be done by constantly repeating specific representations of things, actions, and identities, until what one repeats is naturalized to such an extent that it appears doxic."[70]

The study design of using selected countries of origin as a shorthand for foreigners in general and conceptualizing foreign nationals as temporary work migrants thus supported the long-standing belief that Germany was not a country of immigration, despite overwhelming contradicting empirical evidence. In this sense, structural conservatism and inertia can be regarded quite powerful and productive features of governmental knowledge.

In close connection to the effects of structural conservatism, the BAMF maintains a distinct speaker position throughout the years which can be best described as neutral and objective, the third feature of governmental knowledge. This feature is visible in the impersonal public conduct of Research Group officials: According to the Research Group's head official, the BAMF's researchers do not participate in public or academic debates as individuals, but rather as representatives of the BAMF or the federal government, respec-

68 Ibid. For an overview of state-sponsored knowledge production on return migration, cp. Hönekopp 1987b

69 Boswell 2018, 152ff. For an overview over "Guest Worker" knowledge production, cp. MARPLAN 1995 and Mehrländer 1987.

70 Neumann 2012, 79f.

tively.[71] Objectivity is also a deeply engrained feature of the research content, as evident from the example of the Migration Reports. To an academic reader, the reports with endless repetitions and copy-pasted passages from last year's editions are a rather dry reading experience; at the same time, references to theoretical concepts to explain the conceptual background of knowledge production are constantly absent. Both redundancy and presenting scientific concepts as facts, however, do create a specific perspective of objectivity.[72] This entails an effect of objectification of the research subject through the use of quantitative methods and data, most importantly statistics[73] and legal categorizations.[74] At the same time, the state apparatus itself is objectified, especially its political actions: In the Migration Reports, changes in the legal or institutional make-up are mentioned only in the according year's edition and are never explicated; long-term trends in governance remain largely ignored. The state appears thus as a timeless background unaffected by politics;[75] it is described as an abstract mechanism of legal norms, authorities and policy aims, whose conduct is described in a technical way. Both perspectives precondition each other[76] and join together to a distant, uninvolved view, akin to Haraway's oft-quoted notion of the "gaze from nowhere."[77] However, this perspective cannot be considered a self-explanatory feature of knowledge in a highly politicized policy area such as migration. In fact, the BAMF has a record of publishing alarmist and controversial studies in the past.[78] In contrast to this, the outwardly boring make-up of the Migration Reports and other research publications can be read like a conscious effort of establishing a somewhat neutral speaker position in between alarmist and multicultural positions. In fact, the thrive for objectivity is proportional to the degree of politicization of a given political question:

> "some of our studies draw conclusions on what could be done. But in such a contested area like for example citizenship [...] we didn't do that. We made a proper study, we analyzed [different] effects and presented [the material].

71 Kreienbrink and Worbs 2018, p. 249
72 Doughan and Tzuberi 2018, p. 272
73 Amir-Moazami 2018b
74 Boswell 2018, 1ff.
75 Ferguson 1994, p. 256
76 Cp. Hess 2014, 258f.
77 Haraway 1988
78 Cp. for example Kelek 2006

> If you look at the [...] press releases [of two contesting political actors, V.K.], one could think they referred to two different studies. But [...] we were OK with that, because everyone can work with this material."[79]

This quote illustrates a mechanism through which an objective speaker position is established. It also points to the fact that this speaker position cannot be considered an accidental outcome of bureaucratic styles of text production but rather a conscious strategy. In this sense, objectivity and a distant speaker position can be considered a core feature of the BAMF's governmental knowledge production, especially in highly politicized issues.

The fourth feature of governmental knowledge is partial blindness, especially towards politically irrelevant features of the object of research. Partial blindness as such is of course not an epistemic problem but rather a necessary result of abstraction and categorization: at every step between raw data and the final analysis, some details of the original material are sacrificed in order to gain a clearer picture, more rigidly defined categories, or abstract units of equal magnitude for comparison. This "translation process"[80] and the formation of abstract categories are crucial elements of both scientific analysis and bureaucratic administration.[81] However, this process can turn into a source of bias if consistently the same details are sacrificed in the course of translation; knowledge is then blind against these allegedly unimportant facts which are consistently filtered out. This is what happens at the BAMF as an outcome of the thorough internalization of political relevance considerations: In integration research, for example, negative effects of discrimination on integration success are discussed in accordance with the economic usefulness of a given migrant. Conceptually, the integration of privileged immigrants is portrayed as an outcome of both individual and structural factors, thereby stressing the need of an open society and condemning the negative

---

79 "Es gibt Studien, in denen Schlussfolgerungen drinstehen, was man sinnvollerweise machen kann. [...] Aber in so einem umkämpften Feld wie zum Beispiel [...] Staatsbürgerschaft [...] machen wir das eben nicht. Also haben wir eine saubere Studie gemacht, [...] haben Effekte ausgewertet und haben das dann präsentiert. Und wenn man sich dann die Pressemitteilungen [zweier politischer Akteure, VK] anguckt, könnt man der Auffassung sein, das sind zwei verschiedene Bücher gewesen. Aber [...] damit können wir sehr gut leben, weil dieses Material können alle verwenden."(Interview with a BAMF researcher, 2015)

80 Schiffauer 2018

81 Affolter 2017, 156f., Jenkins, 7ff.

effects of discrimination on integration. This is contrasted by the analysis of most other immigrant groups, which is conducted as if successful integration was solely the product of individual effort. Discrimination for these migrant groups is portrayed as a "perception" or a "feeling"; the BAMF furthermore tentatively suggests that this might be caused by a lack of integration (if, for example, interactions are overhastily interpreted as discriminating due to a lack of proper German skills).[82] In effect, the cause and effect relationship between discrimination and poor integration is reversed, depending on the migrant group and ultimately, the steering rationale behind it. This is an outcome of the fact that the BAMF employs different integration concepts: Large-scale studies on most immigrant groups employ a theoretical model of integration based on Hartmut Esser's integration model. This approach measures integration as convergence of statistical indicators in four categories (cognitive/cultural, emotional, social, and structural integration[83]). Esser's concept is however only partially implemented by the BAMF: Those dimensions which contain individual migrant's features (most importantly language skills, economic situation, etc.) are thoroughly studied, while structural factors (legal barriers, structural discrimination, etc.) are underrepresented. While this selection is usually justified with a lack of according data, it is also caused by an according framework of interpretation: For example, a study on integration course participants revealed that members of visible minorities consistently report the highest levels of discrimination. However, the report does not conclude that this is due to a higher probability of members of visible minorities to be singled out for discriminatory acts. Rather, in line with the individualistic framework of interpretation, the reports suggest (albeit tentatively) that discrimination experience is based on incorrect interpretation of social conflicts as being motivated by racism, which can allegedly overcome by further integration measures.[84] In this example, data is read in a politically useful way to justify "more of the same" integration measures, not in a critical way to fundamentally address the issue of racism towards visible minorities. This hegemonic interpretation of integration as the responsibility of migrants is put into perspective by integration studies of rather privileged status groups such as university graduates, entrepreneurs and self-employed migrants: In this context, discrimination, operationalized as "the feeling of

82 Bundesamt für Migration und Flüchtlinge 2013b, p. 74

83 Bundesamt für Migration und Flüchtlinge 2009c, 207 f.

84 Bundesamt für Migration und Flüchtlinge 2013b, p. 74

being welcome",[85] is reintroduced into the analysis. Here, the Research Group supports the creation of a so-called "Welcome Culture"[86] for example by drawing conclusions for the optimization of bureaucratic procedures for members of privileged status groups.

In sum, the different representation of discrimination shows that integration is conceptualized according to the perceived political use of the knowledge: The Esser-inspired individualistic framework of analysis is applied to legal status groups where governmental intervention is motivated principally by restriction and surveillance.[87] In the context of more privileged migrants whose immigration is supported for economic reasons, successful integration is conceptualized as being dependent on both structural and individual success factors. Again, the different concepts are not primarily caused by a different object of inquiry, or the lack of data, but rather by the standard of political usefulness according to which data and research findings are produced and interpreted. Ultimately, this finding points to the fact that political relevance comes at a cost in terms of scientific independence.

The four features of governmental knowledge – political relevance, structural conservatism, objectivity and partial blindness – arise from the material analyzed in this thesis; while it is reasonable to draw connections to structurally similar cases, they are first and foremost relevant for the institution and time under scrutiny here. Also, it is important to note that the four features of governmental knowledge are not an innate feature of all knowledge produced at the BAMF (or, in extension, other government agencies) but are more visible in some documents than in others. All in all, these features can be regarded rather a departure for further inquiry than an all-embracing list. A too reductionist analysis would not do justice to the BAMF's knowledge production which covers a wide area of topics from regional studies in African, Asian or South Eastern European countries to integration topics such as schooling, media use of migrants, ethnically segregated neighborhoods, as well as descriptive studies of governmental authorities dealing with immigration, integration and asylum.

However, these features are more than just a random list of coincidental findings. Rather, they circle around and focus on one common underlying

---

85 Bundesamt für Migration und Flüchtlinge 2013d

86 Ibid.

87 Bundesamt für Migration und Flüchtlinge 2009a, Bundesamt für Migration und Flüchtlinge 2009c, Bundesamt für Migration und Flüchtlinge 2014b

political structure, namely the nation state: All these features have been analyzed by researchers under the notion of methodological nationalism – the belief that the nation state is the natural container for the social and political reality. [88] Methodological nationalism is an especially salient source of bias in the study of problems which by nature transcend the nation state – such as migration.[89] In this context, this notion has proven to be a powerful source of critique, for example in the study of citizenship,[90] migration sociology and policy making,[91] historical migration research[92] and integration research,[93] among others. For the study of the knowledge production of the BAMF, however, the critique of methodological nationalism has to be qualified: In the case of the BAMF, as a governmental producer of knowledge, methodological nationalism is a necessity, not merely a source of bias. This is again strongly connected to the BAMF's understanding of political relevance: Political relevance is constructed from the perspective of what might be relevant to the national government. This includes a very narrow understanding of *politics* for which the produced knowledge is relevant: Political relevance is increasingly understood as being relevant to the study contractor, i.e. a government agency (usually the Ministry of the Interior), who is usually interested in technical knowledge, not in fundamental critique of its policies.[94] In a similar fashion, the most important mechanism creating blind spots arises from the fact that consistently the same information is filtered and ignored not because it is considered irrelevant per se, but rather irrelevant for the study contractor. Inertia and the corresponding perspective of objectivity both create a specific understanding of the state as a timeless background to the processes under scrutiny. Again, while methodological nationalism is a salient and highly relevant source of critique of knowledge production, it has to be kept in mind that the BAMF has highly profited from this alleged bias: For example, the change in the research strategy from a general sociological research-perspective with according broad focus on society to specific target groups strengthened the Research Group's reputation as a provider of politically relevant knowledge.

88 Wimmer and Glick Schiller 2003, p. 576

89 Beck 2004

90 Hollifield 2004, p. 887

91 Bommes and Thränhardt 2012, p. 202

92 Castles 2000, p. 15

93 Bommes 2009, 130ff.

94 Cp. also Boswell 2009b, p. 174

Research questions are confined to the nation state both in geographical and in conceptual terms again for policy relevance considerations. Administrative data – above all the AZR – is used not because it is especially well-suited, but because it is readily available. Thus, rather than an epistemic flaw, methodological nationalism can be considered the common underlying structure of the four basic features of governmental knowledge production. In the following chapters, this common perspective will be analyzed – not only as an epistemic flaw, but also as a specific logic of knowledge production, distinct and separate from academic knowledge production.

## Research Program

In this thesis, governmental knowledge production will be analyzed mainly in two respects: Firstly, the production conditions of the knowledge, and second, the epistemic features of the knowledge generated.

The analysis of governmental knowledge production at the Federal Office for Migration and Refugees is structured in three parts: Firstly, the history of governmental research on migration will be analyzed. This analysis serves two ends: On the one hand, this history is reconceptualized as a structural precondition of contemporary knowledge production in terms of intellectual traditions and institutional configurations. On the other hand, the framework of analysis for the BAMF's knowledge production sketched out above will be further elaborated using a neoinstitutionalist approach. This approach draws on narratives as a means of conceptualizing mutual influences of policy-making and knowledge production. Narratives are understood as cause-and effect frameworks; in policy-making, narratives are used to reduce complexity and to legitimize political decisions in a given policy area. For analysis, narratives are operationalized with the construction of a specific target group, the identification of key problems, and political solutions to them.[95] This framework is used to analyze the main narratives of past policy-making and research. There are two main sets of sources for this analysis: On the one hand, essays and other research documents issued by the BAMF on the history of migration research, which construct a standard historical narrative of governmental migration research. According to the BAMF, the history of migration research is divided into four eras (refugee/resettler research of the 1950s, "Guest Worker"

95 Boswell 2011

research of the 1960s and 1970s, foreigner research of the 1980s and 1990s and the contemporary era). This narrative will be compared to primary sources, particularly government reports. These reports help reconstructing historical eras of migration research and provide useful material for the comparison to the BAMF's version of this history.

The neoinstitutionalist approach renders the mutual influence of knowledge and policy-making visible: Refugee/resettler research was primarily dominated by a nationalistic frame according to which the newcomers were a priori part of the German nation.[96] This included the claim that resettlers were not migrants and therefore their situation was incomparable to later migration streams – in fact, the inclusion of resettler/refugee research of the 1950s in such a historical perspective is the outcome of relatively recent historical research.[97] In resettler/refugee research, one of the two main research fields was directed towards monitoring various integration parameters (such as employment and income, housing situation and so forth), therefore closely resembling contemporary socioeconomic integration research. The other field of research was directed towards ethnic and cultural aspects which were used to legitimize the inclusion of the resettlers into the national "community of fate"[98] as well as to maintain the territorial claims on lost eastern German provinces.[99] In the BAMF version of history, only the first stream of research is mentioned, since it mirrors the contemporary image of instrumental governmental knowledge most closely.[100]

In "Guest Worker" research,[101] the national paradigm was inverted: The apriori assumption was that migrants were not part of the German nation and would stay only for a limited amount of time. Consequentially, national categorizations were paradigmatic for research, which focused increasingly on "Guest Workers" from the largest countries of origin, as mentioned above. The cause-and-effect arguments reverse around the 1968 recession: While initially, the mutual benefit of the German economy, the migrants and the coun-

96 Bommes 2009, 128ff.

97 Ibid., p. 133

98 Aumüller 2009, 161 ff.

99 Nahm 1959, p. 154

100 The ethno-nationalistic stream of research is not mentioned as such; rather, the apriori community of fate is regarded as a givern (Cp. for example Wollenschläger 2003, p. 41).

101 I follow here Scholten's terminology (Scholten et al. 2015a, p. 319). The BAMF uses sometimes different terms (such as "foreigner research", Cp. Heckmann 2013) to avoid the paternalistic inscriptions of the term "Guest Worker".

tries of origin was stressed, more defensive arguments arise around that time around notions of alternativeless and maintaining the standard of living. This is connected to blaming "Guest Workers" (especially their "culture") for growing problems of the recruitment system. The main elements of this narrative were all constructed during the late 1960s and early 1970s and stay rigidly in place during the next decades, which testifies to the remarkable stability of the "Guest Worker" narrative.

The next phase is sometimes called "a Lost Decade"[102], beginning after the halt to recruitment in 1973, when labor recruitment was stopped due to recession. Until the change of the federal government in 1998, migration policy-making was stalemated by the belief that Germany was not a country of immigration and the successive de-facto inclusion of migrants into social and welfare systems. In this context, the political framing of scientific research according to the "no country of immigration" dogma is very well visible: Integration processes are mostly framed as deficits on the side of the migrants, which are often attributed to culture.[103] Culture serves as the most important explaining variable for the increasing differentiation among the formerly more or less homogenous "Guest Worker" population while increasing political-legal differentiations within the foreigner population are left unregarded. This again follows the political reasoning that administrative measures are a reaction to, not cause of, integration problems. This is remarkable because the BAMF supports a narrative of an increasing antagonism between "rational" science and "irrational" politics[104] which cannot easily be maintained regarding the strong support of the "no country of immigration dogma by governmental knowledge production. This narrative however serves as an important antagonistic picture for the portrayal of contemporary, "enlightened" policy-making, which is portrayed in the instrumentalist picture sketched out above: Policy-making is grounded on sound scientific knowledge and is regularly evaluated and updated according to the latest scientific findings.

In Chapter 3, the institutional make-up of the Research Group is analyzed, spanning from the foundation of the group in 2004/2005 to its contemporary set-up. Again, the core topic of this chapter is to explore the institutional and structural preconditions to knowledge production; the chapter illustrates how

---

102 Bundesamt für Migration und Flüchtlinge 2005a, p. 71

103 Cp. Lanz 2007, p. 82

104 Cp. for example Heckmann 2013, 38f.

the various mutual interconnections between policy-making and science as laid out in Chapter 2 are produced in practice.

The newly established Research Group was confronted with a rather blurry mandate as well as an undefined position in the administrative hierarchy; consequently, the time after establishment can be characterized as a struggle for the establishment of an area of competency and influence. In fact, at least initially, the Research Group can be regarded as a "foreign body" in the BAMF, expressed for example in the professional cultures of officials and researchers, the generational difference between the two groups, and conflicting ideas about the long-term orientation of research originating from the Ministry of the Interior, the BAMF leadership and the Research Group. In the literature, this phase is often used as evidence for the "systematic gap" thesis, according to which "science" and "politics" are systems with fundamentally different functioning logics.[105] This interpretation disregards however the increasing integration of research into governance, visible for example in the fact that more and more studies are commissioned by other government agencies (the Ministry of Interior, above all). This process is interpreted as a strategy of mimicking the function and role of a departmental research institute, with varying success. The result of this strategy is ambiguous: On the one hand, a rising research output and a higher institutional status in terms of staff and resources testify to its success and the political relevance of the knowledge generated. On the other hand, the blurry legal mandate as well as the comparably smaller degree of institutional independence leaves the Research Group in an unfavorable situation in times of institutional conflict especially vis-a-vis other departmental research institutions.

In this process, a distinct concept of governmental research is constructed, which is understood as the practically relevant counterpart to academic research. Researchers characterize their work as practical (in contrast to theory-oriented), flexible (in contrast to methodological rigorism) and pragmatic (in contrast to critical). So far, the often-quoted academic disregard for governmental research[106] seems unjustified: Knowledge production at the BAMF follows the same systematic rules as in academia;[107] empiric data is collected and analyzed with scientific methods, publications are referenced and quoted according to academic criteria. The only difference

105 Kraler and Perchinig 2017, p. 85

106 Cp. for example Kraler and Perchinig 2017, 66f.

107 Barlösius 2008, p. 25

between governmental and academic research, however, lies in the practical relevance criterion, as stated above: research is valued if it is useable in the political arena. This difference is however grave: It causes the above-mentioned structural conservatism and inertia, since study contractors strongly prefer such uncontroversial, mainstream knowledge.

The structural conditions of knowledge production are analyzed mainly with the help of expert interviews and documents. The Research Group has issued a series of articles in which it elaborates something like a mission statement, or what can be termed its "self-understanding". These include articles in scientific magazines, usually commemorating institutional jubilees, and PR material from the BAMF website.[108] This understanding will be analyzed in terms of its capacity as a "double hermeneutics"[109]: that means, not the role of the Research Group as such will be analyzed, but rather its role from the view of the involved actors. This approach stresses the fact that institutions are crucially shaped by the shared beliefs and assumptions of their members, which explain the particular strategic orientation of the institution and its development over time. The resulting shifts in institutional configurations and knowledge production strategies can therefore be linked to according shifts in the self-understanding of the Research Group. Expert interviews of current or former BAMF Research Group members and other relevant specialists in the field constitute the other main type of information source for this chapter.[110] Expert interviews can be considered a standard methodology of anthropology but increasingly gain more currency in political science as well, especially in explorative studies.[111] Despite considerable difficulties of access to the field,[112] nine interviews of two types were conducted: Firstly, off-record informal interviews which were documented in the field notes; and secondly, recorded semi-structured interviews. All interviews were anonymized and all direct quotes have been edited and approved by interviewees.

---

108 Cp. Kerpal 2003, Bundesamt für Migration und Flüchtlinge 2013a, Kreienbrink 2013, Bundesamt für Migration und Flüchtlinge 2015d, Bundesamt für Migration und Flüchtlinge 2015a, Kreienbrink and Worbs 2015, Kreienbrink and Worbs 2018

109 Cp. Walzer 1987

110 See list of interviews in the appendix.

111 Cp. Schneider 2010, p. 32

112 Most interview requests as well as permission to conduct field work during an internship were denied. Quotes form interviews were carefully revised and reformulated by interviewees, and released only under the condition of anonymity.

The remainder of the thesis is dedicated to the analysis of the BAMF's research output in chapter 4 by analyzing the material for specific epistemic features arising from the production conditions as laid out in chapters 2 and 3. In a first step, the research output of the BAMF is analyzed quantitatively. The analysis entails Working Papers, research reports and Migration Reports from 2005 to 2015, spanning from the foundation of the Research Group to relatively recent publications. This data is compared against academic knowledge production to determine the relative size and the selection of topics of the BAMF's research output to its peers in academia. While both academic and governmental research focus on socioeconomic integration, discrimination and multiculturalism, both prominent topics in academic research, are almost completely ignored by the BAMF. Methodologically, a clear focus on quantitative research using data from the AZR is discernible.

In a second step, qualitative literature analysis is conducted. Here, the concept of narratives as well as the governmentality-approach is used to construct selected "knowledge-power-complexes"[113]: This concept is specifically geared towards analyzing the multiple connections between governance and knowledge production by analyzing the specific practice for which the generated knowledge is relevant. As already mentioned, four selected complexes will be scrutinized: Administrative knowledge with the example of the Migration Reports, depoliticizing knowledge in integration research, defensive knowledge to calm the public debate in the context on research on Muslims, and legitimizing knowledge in regional studies of African and Eastern European Migration.

The four types of knowledge-power complexes render an overview over the topics, methods and features of the BAMF's knowledge production: Two chapters cover integration, and two migration topics; Two chapters focus on broad, general research streams (integration and administrative research) while the other two represent rather specialized knowledge on clearly defined research fields (migration potential and Muslims). Finally, the selection covers both theoretical/conceptual aspects of knowledge production (in the case of general integration research and migration potential) as well as the less theory-oriented, data-driven styles of knowledge production (as in the case of Migration Reports).

The thesis concludes with final remarks on the inherent contradiction involved in the production of both politically relevant and objective knowledge.

113 Cp. Mecheril et al. 2013, p. 20

# History of Governmental Migration Research

The BAMF Research Group was founded in 2005, considerably later than any other departmental research institution in the social sciences in Germany. Most of these institutes date back to the 1960s and early 1970s. This is, of course, no coincidence: until 1998, German migration policy-making followed the dogma that "Germany is not a country of immigration." Nevertheless, migration policy and migration knowledge production existed in all but a name before 1998. In this chapter, governmental knowledge production on migration between the end of World War II and the foundation of the BAMF Research Group will be discussed, focusing on the mutual influences of migration policy-making and research.

The aim of this chapter is twofold. On the one hand, it highlights the historical development of the most important key terms and concepts of governmental migration research which collectively constitute the intellectual foundation for migration research today. The mechanisms and practices which shape a specific governmental perspective on migration forms will be discussed here. On the other hand, this chapter analyzes BAMF's representation of this history: as will be demonstrated, BAMF is keen on presenting a picture of the history of migration research consistent with the instrumentalist account of knowledge utilization, a topic also critically discussed.

The analysis is based on findings from documents which can be grouped as two sets of sources. First, there are publications which construct something like an "official historiography" of migration research according to BAMF, or a set of retold statements about the institutional and ideological history of BAMF research.[1] The most important text in this regard is an essay titled "Migration Research in Germany" by Friedrich Heckmann, published in an

1 Cp. Kratzer 2018a for a critical discussion

anthology in 2013 to commemorate the 60$^{th}$ anniversary of BAMF's founding.[2] This official historiography of migration research as established in Heckmann's essay is a point of reference for a later text published for the Research Group's ten-year anniversary in 2015:

> "The research center at the BAMF has to be understood as a part of institutionalized migration and integration research, which is rooted in earlier knowledge production phases of Refugee and Expellee research (until the end of the 1950s) and "Foreigner Research" or "Foreigner Education" (1970s/1980s). [...] The research unit was created in 2005 as a consequence of the paradigm change in German migration and integration policy-making since the turn of the millennium. With the rising acknowledgement of the factually existing situation of immigration and the political will to redesign it, the demand for an institution providing according data and knowledge was created."[3]

As the analysis will demonstrate, the official historiography constructs an image of governmental research which relies on a particular representation of academic migration research, as mentioned in the quote, conceptualized from a perspective of instrumental knowledge utilization. This representation is, however, produced by numerous omissions, ex-post rationalizations, and other inconsistencies, some of which will be analyzed in this chapter.

This analysis will be conducted on the basis of governmental documents from the respective eras, which constitute the second major type of source documents in this chapter. Interestingly, all of the phases of migration research mentioned above – Resettler/Ethnic Germans, "Guest Workers", and "Lost Decade" – coincide with report series on the respective target population: during the 1950s, knowledge on refugees and resettlers was published by the responsible ministry in the *Flüchtlingszählwerk* (refugee registration system); during the 1960s and 1970s, "Guest Worker" research was organized in a report series issued yearly by the Federal Agency for Labor. These reports cumulated in a widely recognized, 1972 representative social survey on migrants which initiated a report series on foreign citizens from the 1970s to the 1990s. While these documents differ in length, topics, and methodology, they share a basic structure of knowledge production: a large part of the reports is

---

2 Heckmann 2013, Kreienbrink 2013, Wollenschläger 2003, Kreienbrink and Worbs 2015, Bommes and Thränhardt 2012

3 Kreienbrink and Worbs 2015, p. 330

dedicated to a description of the population by statistics and numbers, while usually a comparatively smaller part includes multiple studies geared towards legitimizing political decisions. It would, of course, be insufficient to equate governmental knowledge production with the content of these report series; however, for the sake of the analysis, they can be considered a useful source for two reasons: firstly, reports contain administrative data which can be analyzed in terms of how the population group in question is constructed; in addition, arguments about policy – core problems, policies, and arguments directed at the public debate – can be discerned, albeit sometimes quite indirectly. The use, selection, and interpretation of data can then be interpreted in connection with policy relevance considerations to reveal specific governmental perspectives created by the knowledge generated in these reports.[4] Secondly, reporting has a tendency of inertia by default: the genesis of statistical data on a specific sub-group of the population needs to meet rigid scientific criteria of quality. To ensure that a sample of respondents is statistically representative is resource intensive and therefore avoided if possible. This is especially true for heterogeneous populations, such as migrants, where comparatively larger samples are needed to ensure statistical representativeness for sub-groups. Furthermore, data becomes valuable only in comparison to other data, especially if the same information is collected over several years to reveal trends and developments.[5] The resulting inertia is one reason annual reports constitute a prime source for tracing the genesis of a cognitive framework of knowledge production whose features are more clearly visible since stability and continuity is emphasized.

As already mentioned, the basic structure of the official historiography of migration research is made up of three phases (Refugee/Expellee phase, "Guest Worker"/Foreigner research, and Migration and Integration research), each of which has their own institutional set-up, policies, and knowledge produced in that time.[6] With the help of primary documents, the analysis reconstructs the changing institutional and epistemological frameworks of knowledge production over time, identifying the most important systems of policy-making and related knowledge production of a given era. This basic narrative of three phases seems to be a standard description in academic and governmental texts on the history of migration research in Germany; the BAMF Re-

4 Rose 1991, p. 675

5 Research Notes, interview with a government researcher, February 2017

6 Bundesamt für Migration und Flüchtlinge 2013a, 33 f.

search Group has issued several related texts on the history of the institution which are structured in this way.

This history will be analyzed using the concept of policy narratives. Following a neo-institutionalist approach, the basic argument is that policy does not simply emerge directly from objective knowledge. Instead, political actors react to increasingly complicated and unsure situations with the construction of political narratives, mainly in an attempt to reduce complexity, offer legitimization and provide a credible strategy for decision-making. This does not mean that knowledge is only a more subtle expression for propaganda which can be manipulated according to political interest: rather, knowledge is central to these narratives, since they are expected to be firmly grounded on sound empirical facts and must meet rigid scientific standards to maintain the claim of credibility and objectivity.

Referencing Boswell et al. (2011), three major elements of policy narratives will be discussed for each of the respective phases of knowledge production. The first is the construction of a target population, including its size and the main problems connected to it. This question is of paramount importance because it helps to understand the genesis of a variety of status groups in the course of post-war migration which are still today the most important lens through which migration in Germany is discussed politically and scientifically. The second element is the development of a set of claims for the root problems of the phenomenon in question, and third, claims about the question of how policy affects (or has affected) the problem complex.[7]

## Refugee Research

In his 2013 essay on the history of migration research in Germany, Friedrich Heckmann places the beginnings of German migration research from the end of World War II until about ten years later.[8] After the war, millions of people were migrating across Europe for one reason or the other; there were refugees from territories formerly belonging to Germany, German resettlers from Eastern Europe, refugees from the Soviet Occupied Zone, people who lost their homes due to war destruction or expulsion, concentration camp inmates and

7 Boswell et al. 2011, p. 5

8 Heckmann 2013

forced laborers who were brought to Germany during the war, and demobilized soldiers of the German and several ex-enemy armies.

One of the top priorities was the reconstruction of a working administration especially for the newly arrived and dislocated persons. Against the backdrop of the hardships during the immediate post-war period, establishing stable population categories was a challenging task. In respect to the dislocated population, the most important issue was the establishment of clear differentiations between Germans and non-Germans, the latter being for the most part so called Displaced Persons (DPs). These two categories of refugees were clearly separated by institutional organization, legal status, and access to material resources. German refugees were subject to further internal differentiation, most prominently between refugees and expellees, as a definition established by the American Occupation Forces clarifies.[9]

In practice, some easily distinguishable statistical markers like nationality and place of origin serve as the key indicators intended to classify and register a given person quickly and unambiguously into the proper category. The most important factor of differentiation for the Allied Forces was the region of origin of an individual refugee and the resulting degree of permanence of the migration: while refugees were a status group in need of help primarily in order to return home, expellees were regarded as people for whom return was impossible. This in turn justified a more preferential access to material resources to facilitate socio-economic integration. This definition and the hierarchy of statuses attached to it proved to be very stable as demonstrated by the fact that it was used later in German federal law. However, over time, the German administration performed a redefinition of the status hierarchy: the preferential treatment of expellees was interpreted as a compensation for the higher degree of violence and coercion suffered during migration, rather than an integration measure for the permanent stay in the region of destination.

In practice, this registration system proved to be difficult to implement with the statistical data and the administrative structure at hand. Most importantly, many refugees could not be registered because they had already fled before the end of the war; according to estimates, this included about half of the 8 million refugees in West Germany.[10] The most significant statistical marker of citizenship, important for discerning German refugees from

9 Memo by the US occupation forces to the Bavarian council of refugees, 10 April 1946. Quoted after Lemberg and Edding 1959, p. 385

10 Parisius 2003, p. 256

DPs, for example, was in some cases irrelevant, because some expellees were deported from areas that never belonged to the German state (such as Sudeten territories), which meant that these people usually did not possess a German passport to begin with. A similar problem prevailed when using the place of birth or residence as an indicator, since many expellees moved during the war as military personnel or as civilian occupation officials as part of National-Socialist social engineering policies. Another challenge was created by the fact that some status groups were granted access to resources and given preferential treatment, and others not, which made tactical self-declarations more likely. It was therefore important to establish indicators which did not rely only on the information given by the person in question.

To improve the data base of population registration in general and especially refugee classification, a census was carried out in 1946. The fact that this census was one of the very few administrative acts that were executed across all occupied zones speaks to the urgency and importance attached to this measure. Also, a uniform census was helpful in unifying the different terminology, legal prescriptions, and institutional competencies which existed especially between (and in some cases within) the occupation zones.[11] The problems of refugee classification mentioned above were addressed by a new nationality concept in this census, *Volkszugehörigkeit* (ethnic belonging). According to this concept, foreign nationals with proficiency in German and a "commitment to the German People" were regarded as Ethnic Germans.[12] With this concept, a clear differentiation between Germans and non-Germans was facilitated, which in turn enabled the orderly registration and classification of migrants into one category or the other. At this point, a rather problematic effect of administrative continuation should be mentioned: the definition of this new term was not, as contemporary sources suggest, based on administrative rationality alone,[13] but rather the adoption of a slightly reformulated decree by the National-Socialist Ministry of the Interior issued in 1939. For decades later, German administrative courts recognized documents issued on the basis of racist national-socialist policies as such a commitment in the sense of the law; an example of this is registration on the so-called *Volksliste* (ethnic registry) in occupied Poland during World War II.

---

11 Beer 2003, p. 300

12 Bundesministerium für Vertriebene, Flüchtlinge und Kriegsgeschädigte 5/22/1953, § 6

13 Nellner 1959, p. 63

In effect, the differentiation between Germans and non-Germans on the one hand and multiple status groups within the German population on the other has had tremendous structuring effects on the politics of migration. The various administrations built up in the three western occupation zones were later merged into the *Bundesministerium für Vertriebene, Flüchtlinge und Kriegsgeschädigte* (Federal Ministry of Displaced Persons, Refugees and War Victims), while non-German DPs and refugees were registered and cared for mainly by international organizations, such as the Red Cross and several agencies of the newly founded United Nations.[14] Within the German refugee population, various legal groups of refugees were created; those who received the most support were called *Heimatvertriebene* (expellees) from territories that no longer belonged to Germany, such as Eastern Prussia or Silesia, followed by *Flüchtlinge* (refugees) who moved to these territories during the war. *Refugees* from the German Democratic Republic (*Deutsche Demokratische Republik*, GDR) constituted a third, less privileged category.

By 1949, the dramatic after-war period of anarchy was coming to an end. In terms of migration movements, the massive chaotic movement of millions of people was over, since most of the expulsions from former German territories were completed by that date. In the 1950s, the pattern of movement was an orderly stream of east-west migrations of Ethnic Germans. The new arrivals in subsequent years originated either from the GDR or the former German territories in Central and Eastern Europe and were accordingly attributed refugee statuses under the supervision of the Federal Ministry of Expellees, Refugees and War Victims.

In the following years, a parallel development of statistical units and academic knowledge, administrative action, and policy-making around the refugee and resettler population unfolded. Coordinated by the Ministry for Refugees, Displaced Persons and War Victims, a research network of experts from different fields was installed in 1954 which built up most of the body of administrative-statistical as well as academic knowledge usually referred to when resettler research is discussed.[15] This research activity was promoted chiefly out of fear of political radicalization of a marginalized population group which constituted a sizeable portion of the society.[16]

---

14 For an overview of the ministry's history, see Beer 2003

15 Bommes 2009, p. 129

16 Castles and Wihtol de Wenden 2006, p. 233

Resettler research can be separated into two main streams, as mentioned above: one was oriented towards typical population-science type of concepts, research questions, and methods and included economic, demographic, and social issues. The second type was characterized by ethnography, history, and linguistics and covered mostly cultural and historical topics.

The first type of texts contains statistical reports and data-based policy analyses. Concerning the methodology, make-up, and research designs, these studies are relatively similar to contemporary integration studies, a not overly surprising finding given the common roots of both research streams.[17] Many of the authors of these texts are practitioners of the early refugee administration.[18] Most of the methods and concepts stem from demographic research and interpret data from the *Flüchtlingszählwerk* (Refugee Register Mechanism)[19], a dense statistical monitoring system.[20] By 1959, most aspects of the life of the expellees were captured statistically: data on demographics, living conditions, employment, and vocational and academic education among others was collected in short intervals typically ranging from one to three months.

Conceptually, a typical feature of the Refugee Register Mechanism and related social research is the internalization of statistical and administrative concepts, definitions, research questions, and perspectives. This follows from the fact that the statistics mentioned above serve as the main data base for these studies; also, it seems practical to use status groups as a basis for research for the formulation of policy recommendations. For example, a study about the housing situation of expellees presents data about participants of a government housing program.[21] In this way, the categories and statuses, constructed out of rather pragmatic considerations such as the availability of data and the like, become naturalized. The overarching scientific focus, the economic integration of resettlers into the West German society, can be likewise explained. Most of the resettler support programs were installed in order to prevent ethnic or class mobilization and radicalization among the resettler

17 Cp. Angenendt 1992, p. 187

18 Such as Werner Middelmann, a high administrative official in the refugee administration before the founding of the Federal Republic or Peter Paul Nahm, state secretary at the Ministry for Expellees. Cp. Beer 2003, p. 309

19 Nellner 1959, p. 101

20 Middelmann 1959, 276 ff.

21 Cp. Lemberg and Edding 1959, p. 447

population, which was perceived as underprivileged in their access to jobs and housing, as well as other resources vis-a-vis the autochthon population.[22] All in all, in this type of academic literature, refugees and expellees appear collectively as a group, as an orderly registered population whose economic and social integration is closely monitored and whose problems are solved by corresponding governmental measures. Between 1949 and 1969, multiple programs for housing, economic integration, education – together with cash allowances for expellees and refugees – were implemented by the Ministry for Displaced Persons, Refugees, and War Victims. In hindsight, there is a consensus in the literature that these policies have been successful, both in smothering political extremism and in integrating the expellees into the Western German society and economy.[23]

The second stream of academic literature stems from a nationalistic tradition of ethnography and demography, which emerged around the turn of the 20th century. In the academic discourse of the time, the question of the governance of work migration, assimilation, and naturalization was in the center of debate: from the 1870s onwards, Polish and other Jewish migration triggered a debate whether or not they were entitled to German citizenship and what constituted being German in a wider context. This was connected to a growing scientific interest following the political use of the German-speaking minorities in Central and Eastern Europe, which transformed to a discourse of cultural domination and supremacy in the context of a revisionist German foreign policy following the defeat in World War I.[24] A concept of ethnicity was developed around the notion of "German blood" which supported political claims of territory lost in the Versailles treaties.[25] Under National Socialist rule, this concept was used to justify German supremacy in Central and Eastern Europe with a direct link to the most violent expressions of these theories in form of extermination policies during World War II. Arguably, there is a connection between this academic tradition and the legal definition of "ethnic belonging" in the 1953 expellee act, which relies on similar concepts of ethnicity, albeit replacing the racist term "blood" with an essentialist understanding of "culture." With this background in mind, a strong case can be made

22 Bommes 2009, p. 129

23 Heckmann 2013, p. 35

24 Jureit 2012, p. 26

25 Aumüller 2009, 161 ff.

against the narrative of commissioned research to support bio-political measures; rather, it seems that a highly ideological tradition of ethnic research has successfully adapted to new political realities and continued to influence politics and law-making.

Expellees as an object of knowledge are thus formed by two distinct traditions of knowledge production. While population research was designed to monitor socio-economic indicators, ethnographic research has had an important political impact on the discursive framework of political legitimization. Both levels of knowledge production are visible in the aims of the refugee policies: contemporary sources state quite openly that the political pacification of the refugees and expellees by means of economic integration is one of the most important policy goals, which is then carried out by a system of policy-making and scientific monitoring mostly in different socio-economic fields. However, the aim of economic integration has been contextualized as a policy of burden-sharing and a "compensation for war victims" which was supported by the discourse provided by the ethnic stream of refugee research. One of the most important and extensive works in this research tradition was a collection of crimes committed against the German expellees which served as a justification for the material compensations to this group. Interestingly, in this context, arguments are being brought forward *against* cultural integration, as in the following quote of Peter Nahm, the long-standing state-secretary at the Ministry for Refugees:

> "Not only the Soviet Zone Refugee, but the expellee as well is a full citizen of the Federal Republic representing all of Germany; he does not become assimilated Bavarian or Hessian, but stays Silesian, Eastern Prussian or Pomeranian. This is why the Federal Republic also represents the Eastern Provinces, whose administration has been appropriated by Poland and the USSR."[26]

In the quote, a geo-political dimension of culture becomes visible: in the Cold War era, expellees were one of the most important discursive foundations to the claim of regaining the eastern territories lost to Poland and the USSR. Another trace of this idea is visible in the incentives for expellees to work in agriculture so a sizeable portion of the expellee population could do agricultural work and thereby facilitate the future repopulation of eastern, predom-

26 Nahm 1959, p. 154

inantly rural territories after their eventual annexation.[27] Maybe the latter is not a significant example in terms of political impact, but it is an example of a knowledge informed policy which does not fit the understanding of migration research as interpreted in the official historiography of BAMF.

In conclusion, the official historiography of BAMF offers a quite accurate image of the population-science stream of knowledge production on Ethnic Germans and resettlers. However, it does not mention the ethnic-historic research traditions despite their visible, albeit declining, influence on policy-making. It seems that this selective representation is influenced by two factors: first, the ethnographic research on resettlers does not fit the image of "commissioned research" since it stems from older traditions of nationalistic knowledge production; second, the overt political character of research, as demonstrated in the analysis, deviates from the somewhat apolitical concept of science as a source of technical information for policy-makers. In abstract terms, the BAMF history directly refers to the administrative stream of knowledge production, while the symbolic knowledge produced in this context remains invisible.

## "Guest Worker" and "Foreigner" Research

The next phase of governmental migration research began in the 1960s and is connected to the recruitment of "Guest Workers" from 1955 onwards. After the immediate post-war period, especially after 1961 when the influx of immigrants from the GDR was coming to a halt, economic scientists predicted a serious shortage of labor which could not be satisfied domestically. Thus, a series of bilateral contracts between Germany and several Mediterranean countries established the basis for a large-scale international job placement system which constituted the main channel of immigration into Germany until 1973.

"Guest Worker" recruitment was organized by several authorities in the area of responsibility of the Federal Ministry for Labor. Employers commissioned these authorities with the mass recruitment of a fixed number of workers against the payment of a fee calculated on a per-head basis; the labor authorities were then responsible for the selection, recruitment, and transport of the workers to Germany. For this task, the labor authority set up regional

27 Cp. Bundesministerium für Vertriebene, Flüchtlinge und Kriegsgeschädigte 5/22/1953

offices in several Mediterranean countries to conduct job placements, medical exams, and organized transports.

In the official historiography, research on "Guest Workers" is characterized by academic disinterest which gave way to increasing efforts of knowledge production only after 1973, when migration movements translated into more stable living arrangements. According to the BAMF, one exception to this general trend of disinterest is macro-economic studies, which consisted of cost-benefit calculations setting off the economic gains from foreign employment against infrastructure costs.[28] This argumentation follows a common conception in academia according to which migration was not considered a topic worthy of scientific enquiry, and the "Guest Worker" system, as the name implied, ensured by a strict rotation principle that the presence of foreigners was a temporary phenomenon.[29] Both of these assumptions are, however, disproved by recent historical research on the topic. According to government documents from that era, the "Guest Worker" system was never designed to ensure strict worker rotation; the administration already considered the permanent settlement of "Guest Workers" a *fait accompli* by the early 1960s. Second, especially within the labor administration, an elaborate documentation and reporting system accompanied the increasing recruitment activities; the most important documents in this regard include a yearly report series starting in 1961 with the most important statistical and administrative information on foreign employment as well as several representative surveys. While this literature was omitted in the BAMF historiography, these sources are useful to trace the emergence of a specific framework of scientific analysis of migration which became hegemonic for decades to come. Most basic principles of 1970s and 1980s Foreigners Research – for example, the strong emphasis on employment, the method of constructing and comparing national groups, and data collection by social research as well as administrative registries – were essentially developed in the heyday of "Guest Worker" recruitment.

Who is a "Guest Worker" according to these reports? The definition of the target population is surprisingly blurry from a legal perspective, since the "Guest Worker" system relied on a multitude of legal instruments for work migration which were furthermore subject to a gradual change over time. Furthermore, the term "Guest Worker" is officially avoided until the beginning of

28 Heckmann 2013, p. 35

29 Eg. Wilpert 1984, p. 307

the 1970s; instead, terms like "foreign laborers from recruitment countries" or similar expressions are used.[30] Despite this changing vocabulary, the reports visibly refer to a unified group of migrants whose common denominator is the notion of foreignness and their economic function as laborers. "Guest Workers" are understood as all migrants from "the recruitment countries," giving the impression of a planned, administratively steered recruitment process. A formal recruitment process was, however, by no means mandatory for all work migrants from the countries in question here; alternative forms of job placement and migration increased over time.

In the Federal Agency's reports, "Guest Worker" employment is discussed in reference to foreign employment in general and is listed along with other work migration forms such as migration from neighboring countries and migrants from within the European Community. This framework suggests a legal order of migration channels as well as a relative balance between the various forms of immigration for work purposes. Over the years, the reporting displays a general trend to establish two groups of foreign employees: on the one hand, the aforementioned "nationals from recruitment countries," and on the other, migrants from neighboring countries and member countries of the European Communities. To justify this method especially in regard to the very diverse second group, the Federal Agency for Labor refers to publicly perceived foreignness in combination with large immigration numbers as a principle adopted in its analysis: here, the argument goes that "other" foreigners, with Austrians as a prime example, are not regarded as foreign, whereas "Italians, Spaniards, Greeks and Turks make up a large share of all foreign employees and are thus regarded as typical foreigners by the public."[31] In the 1972 representative survey, all European migrants, pendular migrants as well as recruited workers from numerically less important recruitment countries such as Tunisia, Morocco and Portugal, are similarly grouped together in the "other" category. "Guest Workers" are all Italian, Spanish, Turkish and Greek nationals, irrespective of their actual migration status.[32]

All in all, the formation of two distinct features of governmental migration research can be traced back to the report series issued by the Federal Agency for Labor: first, the method of comparison between national groups, and second, the focus on work migration. The first item is visible in the logic

30 Bundesanstalt für Arbeit 1964, p. 22, Schönwälder 2003, p. 138

31 Bundesanstalt für Arbeit 1964, p. 7

32 Bundesanstalt für Arbeit 1973, p. 15

of comparison according to national groups within the "Guest Worker" population. This order of knowledge reveals what is of most interest in this context: the differences between nationality groups of "Guest Workers" which are perceived as paradigmatic others. This structure overrides both internal divisions within national groups (for example, according to the legal status, which varied by nationality, length of stay, and other factors) and the growing diversity of the "other" group (citizens of EC-member states, numerically less important recruitment country nationals, refugees). The second point refers to a methodological flip in conceptualizing the "Guest Worker" population: while initially, "Guest Workers" were defined as work migrants from a specific set of recruitment countries, in the 1972 representative survey, *all* nationals from these countries were considered "Guest Workers." Again, methodological reasons can be found, but this changed notion also marks the normalization of the concept of "Guest Worker." All in all, as is evident from the government report series on foreign laborers, knowledge on "Guest Workers" focuses on two main features: first, knowledge is constructed around an understanding of inherent foreignness; second, "Guest Workers" are considered an essentially homogenous population of work migrants, which is, for example, visible in the practice of merging different legal status groups and migration practices. This reflects a trend in the reports to essentially equate "Guest Worker" with foreigners in general and single out this particular social phenomenon in terms of analysis, discussion and problematization.

## Policy Legitimization

A growing section of the yearly reports on foreign employment is dedicated to discussions on the advantages and problems of the recruitment policy. An analysis of these arguments displays a distinct shift in the discursive strategy of legitimization from a rather optimistic, opportunity-oriented reporting to a rather defensive, risk-avoiding style of argumentation.

By the beginning of the 1960s, the recruitment of unskilled workers had changed in regard to the employment structure: increasingly, recruitment shifted from seasonal jobs in agriculture to permanent employment in industry. This change can be explained by the ongoing boom in the labor market which resulted in full employment since 1960; in this context, work migration is increasingly conceptualized as a strategy to counter shortages of labor.[33] In

33 Bundesanstalt für Arbeit 1972, 3f.

the years before the 1967 recession, the reports emphasize the mutual benefit for all involved parties – sending countries, the German economy, and to a smaller degree the migrants themselves. The argument goes that the German industry is able to satisfy the demand for labor through recruitment especially of un- and semiskilled laborers. Pre-established work postings ensured that the supply with "Guest Workers" is deployed precisely where the demand for labor is greatest; "Guest Workers" are described as a highly flexible workforce in terms of qualification, location, and economic sector of the occupation.[34] At the same time, costs in social infrastructure like schools, housing, and similar could be kept to a bare minimum: most "Guest Workers" were recruited as individuals, so that practically no children and only to a small extent non-working spouses had to be accommodated. Furthermore, migrants were expected to live in designated collective accommodation; in fact, the provision of such accommodation by employers was a legal prerequisite to employment.[35] According to the reports, the governments of the countries of origin similarly profited from recruitment through unemployment reduction and regional development; in this sense, recruitment was a remedy to structural unemployment especially in rural, less developed areas and among unskilled workers. In fact, over time, recruitment patterns confirm that placement activities shifted from central regions usually in the vicinity of the recruitment offices in the capital to rural areas. Finally, returning workers were expected to contribute to the development of the sending country's economy through remittances and, after eventual return, a transfer of knowledge from the highly developed German industry. Paradigmatic in this respect is the 1965 report on foreign employment which lists several economic, social, and financial advantages for the sending countries to conclude that the recruitment system constitutes an "indirect development aid."[36]

While especially in the beginning of the 1960s the last argument (development by return) was emphasized, these overly optimistic expectations were gradually replaced by a monetary argument, emphasizing the effect of remittances both for the migrant's family and for the sending country's economy in general.[37] In fact, the Federal Agency devotes considerable research resources to determine the amount of money transferred abroad by regularly quoting

34 Bundesanstalt für Arbeit 1971, p. 4

35 Bundesanstalt für Arbeitvermittlung und Arbeitslosenversicherung 1962, p. 12

36 Bundesanstalt für Arbeit 1965, p. 5

37 Eg. Bundesanstalt für Arbeit 1971, p. 5

estimates of the German Federal Bank; related questionnaire items are part of the 1968 and 1972 representative surveys and become a standard item of migration research after that.[38]

In the early phases of recruitment, especially technical problems associated with "Guest Worker" recruitment are discussed, such as housing, transportation, and worker fluctuation. In this context, a paternalistic, sometimes openly racist perspective emerges:

> "Workers are given all the important information orally. This is important because the major part of the recruited workers is unable to process written information, even in the simplest language. Individual workers state over and over again that they have not been informed properly; this is in most cases not the result of ill will, but rather [...] of the lack of the ability to process information correctly. In the future it is important to [...] inform these persons more adequately (with audio tapes, slide shows, etc)."[39]

In a similar way, housing conditions – one of the most pressing problems in the early phases of recruitment – are discussed; worker housing had to be provided by the employer who often relied on barracks, temporary structures and inadequate housing to cut costs. At the same time, employers are presented as benevolent partners engaged in problem-solving; migrants, however, are described as inadequate and in need of supervision:

> "Unfortunately, it has to be stressed that many foreign workers lack the necessary discipline and cleanliness; especially in staff accommodations without supervision. [...] The inclination of workers to move out of even the most exquisite collective housing into private flats has further increased."[40]

In conclusion, the argumentative structure of early "Guest Worker" research is dominated by a framework of reference to cyclical market forces smoothed over by the recruitment program to mitigate the negative effects, particularly labor shortage, of a liberal economic policy. "Guest Workers" are conceptualized in this context as a highly flexible workforce at the disposition of the administration; recruitment is presented as an essentially self-steered process in reaction to the ups and downs of the economy. "Guest Workers" emerge as true economic beings in this context, since most of their behavior is explained

38 Bundesanstalt für Arbeit 1972, 5f.

39 Bundesanstalt für Arbeitvermittlung und Arbeitslosenversicherung 1962, p. 12

40 Bundesanstalt für Arbeit 1965, p. 9

as a passive reaction to market forces or administrative measures. This perspective assesses any behavior as negative which does not fit the assumptions of this framework: many of the problems discussed – housing, lack of understanding, fluctuation – are usually presented as inadequacies of character or education of the migrants.[41]

This general leitmotif of "Guest Workers" as passive recipients of policy measures is also discernible in the argumentative structure of explaining a general trend of increasing migration independence from the mid-1960s onwards. In describing the organization of worker immigration, the reports note the growing importance of migration paths outside of the recruitment system, most importantly personal invitations, self-organized migrations ("second way") or ex-post legalization of immigrants without a work visa (e.g. after immigration on a tourist visa). Especially "second way" migrations were of notable importance in terms of volume by taking advantage of the privileged possibility to immigrate from the recruitment countries with a work visa independently of placement management of the recruitment commissions. This immigration channel became increasingly popular over time as established migration channels provided the necessary information and organization of transport, accommodation, paper work, and job offers outside the control of labor authorities. Personal invitations were another form of immigration whose popularity rose over time: the share of personal invitations reached 45% of all placements in 1972.[42] Invitations were issued to recommended persons via trusted "Guest Workers" who usually selected candidates among relatives or acquaintances, thus creating chain-migration networks between communities in recruitment countries and certain employers or regions. Both invitations and "second way" immigrations gained relative importance over time, so that in 1970, only a minority of about 42% of all incoming "Guest Workers" was in fact recruited.[43] The fact that this increasing independence of migration processes was left unnoticed was not due to a lack of data; rather, it can be argued that this independence was

41 Piore 1979

42 Bundesanstalt für Arbeit 1972, p. 19. In contrast to the aforementioned "second way," the invitations were managed by the recruitment administration, so that all of the necessary steps – registration, medical check, transport – were the same as for anonymously recruited workers.

43 Cp. Bundesanstalt für Arbeit 1972, Bundesanstalt für Arbeit 1971

at odds with the image of a passive, clueless migrant and was therefore disregarded.

From 1970 onwards, the authorities became increasingly wary of alternative migration channels. While personal invitations were initially praised as a method to curb excessive worker fluctuation, legal prescriptions for personal invitations were increasingly tightened. Similarly, "second way" immigration, originally valued as a source for well-trained workers, was abolished altogether in November 1972. However, the increasing popularity of immigration outside the recruitment agencies was never conceptualized as an indicator of an increasing independence of the migration system as a whole. Rather, the lack of control over migration was presented as a problem which could be simply "turned off" once the administrative prescriptions were changed.

The argumentative structure in regard to the legitimization of "Guest Worker" recruitment shifts over time and can be characterized by a gradual retreat to more defensive, technical, and apolitical positions. This is evident in the reports after the recession of 1967: increasingly, the reasoning centers less on the mutual profit for all involved parties and the natural flow of economic tides. Rather, the notion of a permanent foreign worker population as a structural feature of the labor market was increasingly stressed. In general, the argument is no longer that everyone profits from work migration; rather, that there is no viable economic alternative to it. The Federal Agency draws a picture of a modern lifestyle which stunts the maintenance of economic growth for socio-cultural and demographic reasons: the population is aging, meaning less and less people enter the labor market annually. Furthermore, the modern lifestyle developed during the booming era of economic growth is identified as one core reason for the necessity of recruitment:

> "The aspiration for additional free time combined with shorter weekly working hours [...] are factors which will further reduce the work volume of the population. On the other hand, it is not plausible to assume that technical progress of the economy will allow for a reduced labor force. That means that this labor gap will have to be filled by foreign workers."[44]

This analysis is accompanied by historical comparisons to immigration in the years before World War I, suggesting that large-scale foreign employment is not unprecedented and has in fact been a structural feature of the economy

44 Bundesanstalt für Arbeit 1972

for a long time.[45] It is interesting that the obvious historical predecessor of National Socialist foreign labor schemes is omitted in this context. In fact, the practical implementation of "Guest Worker" recruitment was influenced by labor schemes developed under National-Socialist rule.[46] Especially in the case of Italian "Guest Workers," a clear historic continuation of recruitment practices and personnel, as well as employment possibilities (for example at the *Volkswagen* factories in Wolfsburg) can be retraced. While these policies are omitted, the reports contain a reference to international work migration before World War I:

> "Foreign employment is a stable, almost invaluable part of the economy for a long time. Even before the turn of the century (June 1895), more than 315,000 foreign workers were employed in the German Empire. [...] After the Second World War and the decline of war-related unemployment, foreign labor regained its significance."[47]

In this quote, "Guest Worker" employment is presented as the historical rule, not the exception. This difference to earlier reports is quite remarkable: "Guest Worker" employment is no longer conceptualized as a planned political strategy, but rather a quite natural and alternativeless phenomenon.

## Conclusion

The development of the knowledge produced in connection with "Guest Worker" recruitment can be characterized as a general broadening of the research perspective. While early reports focus exclusively on economic aspects of foreign employment, social and cultural factors gain more and more attention over time. The Federal Agency explains this shift with the general expansion of the volume of "Guest Worker" employment and the resulting demand for statistical data. At the same time, information is provided to "support for decisions" to political actors and prepare the background for a "fact-based public debate."[48] Both these uses – instrumental knowledge for policy-making and for calming the public debate which is perceived as "emotional," "irrational," or misled by misunderstandings – are classic topics

45 Bundesanstalt für Arbeit 1965, 3f.

46 Wilpert 1984, p. 306

47 Bundesanstalt für Arbeit 1972

48 Bundesanstalt für Arbeit 1973, p. 10

of knowledge production in a governance context; "clear thinking"[49] – using objective information, rational argumentation, and expert knowledge – is presented as a remedy against the unjustified or unsubstantiated critique of the public and at the same time provides the basis for rational decision-making in the political process.

All of this has some implications for the assessment of the recruitment system in recent historical migration research. In hindsight, the "Guest Worker" system was portrayed as a rotation system whose efficiency was undermined by sluggish bureaucracy, uncooperative employers, or deviant migrants.[50] However, as the analysis of legitimizing knowledge of the "Guest Worker" system shows, rotation has neither been the practice nor the theoretic ideal of the "Guest Worker" system. Since the mid-1960s, the main concern was to attract more workers; in this context, the Federal Agency openly positioned itself against a rotational principle, which would inhibit recruitment especially in the case of skilled personnel.[51]

The gradual prolongation of work contracts was documented quite closely but not discussed as a problem in its own right. On the contrary, the Federal Agency considered the excess fluctuation of workers as one of the core problems, together with growing difficulties in finding qualified workers from the mid-1960s onwards; the Agency attempted to ease both problems with family reunification.[52] By 1970, the Federal Agency considered foreign employment a permanent feature of the labor market, so that at least a share of the workers who were not only seasonally employed was destined to stay for a longer time. Again, this indicates that long-term settlement was not a result of policy failure, or an unintended outcome but rather consciously fostered by the administration.[53] The "Guest Worker" system relied on long-term settlement of

49 Cp. Straubhaar 2003, p. 122

50 Cp. for example Wollenschläger 2003, 41f. This argument is supported by the legal prescriptions of the foreigner's law, according to which in principle every work permit could be renewed annually only on the condition that no German worker was available for the job in question. A gradual tightening of the conditions, especially during the 1967 recession, has been interpreted as a growing inclination to implement the rotation system more efficiently, even if authorities seldom made use of the possibility to end a work contract against the will of the employer. Cp. Dohse 1981, p. 323

51 Bundesanstalt für Arbeit 1974, p. 6

52 Castles 2000, p. 47

53 Schönwälder 2003, p. 125

workers rather than seasonal rotation, based on the assumption that foreign employment was a structural feature of the labor market.

## A "Lost Decade"

In the wake of a recession in 1973, the Minister for Labor issued a halt to foreign labor recruitment. The general belief was at that time that as a result, the foreign population would somehow automatically disappear. According to a 1974 prediction, it was expected that the emigration rate of about 15 to 20% annually would reduce the foreign worker population within a few years to very low levels.[54]

Despite these predictions, the halt to recruitment did not result in a shrinking foreigner population as expected. Although there was a measurable effect in curbing the employment of foreigners, the total number of foreigners did not diminish in the long run. In the years from 1973-1976, the immigration of foreign nationals dropped by more than 50%. From 1976 onwards, however, the migration rate increased due to rising numbers especially in family reunification and, increasingly, asylum migration.[55]

The Federal Government reacted to the unexpected resilience of the foreign population with a bundle of short-term measures. In 1975, the government raised child-support benefits, but extended these benefits only to children living in Germany. This triggered a wave of immigration of children previously living in their countries of origin. Another political measure with detrimental effects was the "deadline date" which prevented family members who immigrated after November 30th, 1974 from obtaining a work permit. The intention behind this measure was to render family reunification economically less attractive and, ultimately, force immigrants out of the country; however, as a result, "many of the young people concerned were unable or unwilling to return to their country of origin, and remained in Germany as 'non-persons' – entitled neither to work nor social-security benefits."[56]

Overall, the years after the halt to recruitment were characterized by a rather inconsistent reaction to the growing independence of migration movements together with increasing costs and problems in connection with the

54 Bundesanstalt für Arbeit 1972, p. 3

55 Numbers quoted after Statistisches Bundesamt 2017, p. 69

56 Castles 2000, 48f.

social infrastructure, particularly housing and education. Policy-making was confined to classic instruments of "Guest Worker" recruitment which were developed in times of an expanding labor-market and mostly tailored to satisfy a steadily growing demand for unskilled labor. Through the halt to recruitment, the authorities lost the only active migration steering mechanism at their disposal. This policy failure coincided with an economic recession which dramatically reduced labor demand in unskilled industrial jobs as well as in the mining industries.

In 1978, the Federal Government created the office of "Commissioner for Foreigners", whose first representative Heinz Kühn presented a widely quoted report on immigration and integration in 1979.[57] In this report, Kühn called for the official acknowledgement of a sedentary foreign population and social and economic integration policies. In his view, these should target especially the "second generation" which was identified as the most problematic social group in this respect. In the same year, a commission consisting of the most important political actors offered policy guidelines to the Federal Government, largely rejecting Kühn's proposals and proposing a rather authoritarian stance towards immigrant integration to deter additional in-movements. The two documents represent the poles of policy-making during the "Lost Decade" which was characterized by a political stalemate: on the one hand, the most important policy guideline was the belief that Germany was "no immigration country" – a programmatic principle repeated over and over "like a dogma"[58] – but on the other hand, increasing criticism targeted the obvious incoherence of this dogma in the face of a steadily growing need for pragmatic policy interventions on a local level. Institutionally, this stalemate divided the political system into two large camps: the "No Immigration" line represented by the two conservative parties (CSU and a majority of CDU) and the Ministry of the Interior versus the more progressive foreigner policy in general – expressed in various political claims of foreigner integration, social policy, and granting of political rights – supported by a large coalition of interest groups, worker unions, the Ministry of Labor, the Foreigner's Commissioner, the Liberal, Social-Democratic and Green parties, and churches.[59] This constellation is in some respects the most important point of symbolic reference for contemporary migration-policy making to the extent it serves

57 Kühn 1979

58 Bade 2016, p. 53

59 Herbert 2000, p. 278

as a negative example. In hindsight, political leaders and the BAMF alike refer to the 1980s and 1990s as a period of "stalemate" characterized by a "lack of policy coherence" and signifying especially in regard to neglected integration policy a "lost decade."[60] The term was coined by historian Klaus Bade in the influential 1993 "Manifesto of the 60."[61]

Against this backdrop of political deadlock, knowledge production on migration shifted during the second half of the 1970s. This shift is characterized by a growing volume and diversity of academic research, but also a growing commitment of state research institutions to migration research. In a survey on migration-related research projects between 1975 and 1989, Angenendt (1992) concluded that roughly two-thirds of all research projects were carried out at universities, while the remainder was about equally divided between state and private institutions.[62] While research in this period is characterized by a growing diversity of topics, almost all research projects can be considered socio-scientific and are either foundational data collections or application-oriented studies of specific target groups (for example, "Second Generation", Women, Return Migrants). Governmental knowledge production shifted its form and focus as well: instead of a yearly report on foreign employment, a report series of studies based on the 1972 representative survey was established with new issues roughly every five years with a largely unchanged methodology and catalogue of research questions.[63]

The framework of knowledge production on foreigners as inherited from the "Guest Worker" era continued to serve as the main template for governmental knowledge production, while ad-hoc additions were made to accommodate demographic and legal changes. As mentioned, during the 1960s, a method of reporting was established which equated foreigners more or less directly with "Guest Workers" who were seen as a uniform group of "real aliens" or "newcomers."

In line with the standard principles of data collecting by government authorities, the reports name mostly methodological reasons for continuing the research paradigm, most importantly the foundation of time series to make data comparable over a long period of time. Consequently, the framework

60 Bundesamt für Migration und Flüchtlinge 2005a, p. 71

61 Bade 1994

62 Angenendt 1992, 181ff.

63 Bundesministerium für Arbeit und Sozialordnung 1996, p. 22

of analysis, the methodology of research, the focus on work- and return related questions remained to a large degree unchanged although it became increasingly clear that minor adjustments were insufficient to reconcile the "Guest Worker" framework with the empiric realities. For example, the base population was changed in 1995 to exclude Spanish and Greek nationals, as well as migrants living in East Germany, due to their dwindling numbers. In the same year, Polish and Vietnamese migrants were added to the report to accommodate post-reunification migration movements and GDR "Guest Worker" migration, respectively.[64] Despite these changes, the reports document the growing difficulties of the "Guest Worker" framework of analysis to capture the increasing diversity of migration flows. These difficulties are the outcome of three interlocked processes which diversified the hitherto relatively homogenous migrant population in three respects: Firstly, socio-demographically, secondly, ethnically, and thirdly, in legal terms.

Referring to the first point, in terms of working arrangements, the foreigner population fulfilled to an increasingly shrinking degree the demographic and economic characteristics of "Guest Workers." This change affected most of the specific demographic and socio-economic features arising from the "Guest Worker" system such as a high percentage of young people, the high employment rate, and a surplus of men. While in the mid-1960s the average labor market participation amounted to 65%, this rate dropped to 52% in 1989.[65] By 1987, almost half of the foreign population which was usually referred to as "foreign workers and their family members" was in fact not employed. Furthermore, foreigners were increasingly less likely to be employed in un- and semiskilled labor in the industry; work arrangements which were habitually associated with "Guest Workers". At the same time, the share of skilled workers, employees, and self-employed foreigners rose steadily. In effect, an ever-decreasing number of foreign nationals fulfilled the various socio-economic criteria of "Guest Workers": by 1986, only 55% of employed foreigners possessed what was commonly associated with a "Guest Worker job," and this share decreased further to 39% in 1995. Self-employed working arrangements, by definition excluded from governmental reports, increased almost fourfold in the same period of time to 9%.[66]

64 Bundesministerium für Arbeit und Sozialordnung 1996, p. 1

65 Bundesministerium des Innern 1989, p. 7

66 MARPLAN 1995, p. 6

Secondly, the inflow of migrants diversified in terms of the countries of origin. During the "Lost Decade," asylum and family reunifications replaced the recruitment system as the main channels of immigration. In reference to the former, during the 1980s, the influx of asylum seekers amounted to 30,000 to 50,000 people annually, an amount with a rising tendency towards the end of the decade.[67] For the method of knowledge production by comparing "Guest Worker" nationalities, this influx was challenging: the increasing diversity of migration flows challenged the assumption of equal legal treatment to some degree. Furthermore, the diversification of countries of origin rendered the method less representative of the foreigner population in general. While "Guest Workers" from Italy, Turkey, Greece, and Yugoslavia comprised 82% of the foreign population in 1972, their share decreased to 59% in 2001; the base population represented in the surveys is still smaller due to the exclusion of self-employed migrants, asylum seekers, and, since 1995, migrants living in East Germany.[68]

Thirdly, in a related process, the foreigner population became more and more diversified in terms of legal statuses. During the "Guest Worker" era, most foreigners possessed similar work and residence permits; the standard method of comparison across national groups could therefore rightfully assume a level legal playing field. However, this level field shifted: in the aftermath of the halt to recruitment and during the 1980s, a process of political steering for a number of legal regulation complexes evolved, targeting family reunification, asylum, and residence permits.[69] Administrative measures were redesigned to selectively curb the inflow of migrants, usually by limiting incentives to immigrate in combination with strategies of deterrence, especially for asylum seekers, but also in the area of family reunification.[70] Judicial protection of migrants' basic rights from overly aggressive expulsion and rejection policies carved out increasing social, economic and civic rights, but these rights were applied selectively most importantly for long-term stayers, which further increased the diversity of legal status groups according to the duration of stay in the country. At the same time, European legislation

---

67 Bundesministerium des Innern 1989, p. 44

68 Bundesanstalt für Arbeit 1974, 9f. and Bundesministerium für Arbeit und Sozialordnung 2001, p. 6

69 Gusy and Müller 2012, p. 9

70 Joppke 2001, p. 48 mentions three paradigmatic decisions by the Federal Constitutional Court in these areas of legislation.

ensured unlimited work- and residence permits for Italian, Greek and other European nationals. As a result, a 1989 report of the Ministry of the Interior lists five residence and four work permit classes, different types of family reunification schemes (depending of federal country legislation), as well as four different classes of refugees (asylum applicants, Eastern Bloc refugees, contingent refugees, and "de-facto" refugees).[71] For the "Guest Worker" framework of analysis, the most problematic aspect of this legal diversification was the fact that this hierarchy of legal titles was not distributed equally across the "Guest Worker" nations. Rather, by virtue of long-term settlement or EC-legislation, respectively, Italian, and later Spanish and Portuguese nationals were more likely to possess unlimited residence and work permits while Yugoslavian and Turkish "Guest Workers" were much more likely to be affected by tightened legal provisions in respect to work permits and family reunifications. In 1980, the majority of Turkish and Yugoslavian residents had to renew work- and residence permits at least every two years, while all Italian nationals possessed permanent residence and work permits due to EC regulations; Spanish, Greek and Portuguese nationals were more likely to possess unlimited residence titles across all age groups than the average foreign national.[72] The legal inequality across "Guest Worker" groups was further increased by the fact that both Yugoslavian and Turkish national groups accommodated an increasing share of refugees from the 1980s onwards, which rendered these two groups more heterogeneous and less privileged in comparison to Italian, Spanish, and Greek foreign nationals.

As a result of socio-economic, ethnic and legal diversification, the relatively rigid framework of comparison between "Guest Worker" nations became gradually less meaningful and representative during the "Lost Decade" despite the considerable methodological effort put into the governmental reports.[73] The concept of comparison of the larger national groups, as well as the focus on semi-and unskilled laborers, covered an increasingly shrinking share of the migrant population and migration-related social phenomena. At the same time, the concept was blind to the increasing internal diversification of the respective national groups as a result of selective effects of restrictive

71 Bundesministerium des Innern 1989, 11ff.

72 Der Bundesminister für Arbeit und Sozialordnung 1981, 565ff.

73 For the representative surveys, more than 1000 interviews per national group were conducted.

regulations and refugee migration. This, in turn, rendered the concept of national groups less meaningful as essentially equal units of comparison.

Accordingly, the BAMF-history of migration research sweeps rather briefly over this period and mentions foreigner education as the only example of applied governmental research, accompanied by a growing body of academic research which becomes more and more independent through the development of theory and the establishment of dedicated academic migration research institutes.[74] However, it is incorrect to assume that governmental knowledge production during the "Lost Decade" ceased to exist; instead, it centered on specific sub-groups of the general foreign population which were perceived as especially problematic or otherwise qualified for increased political and scientific scrutiny. One important aspect in this context is the shift of policy authority from the Ministry of Labor to the Ministry of the Interior. This process was gradual and incomplete, but it can be connected to a general recalibration of foreigner's policy to principles of law and order instead of social and economic policy in accordance with the "no country of immigration" dogma.[75] In terms of knowledge production, this led to the conceptualization of foreigners as a potential threat to public order and the introduction of a security-oriented governmentality logic which focuses on target groups that are perceived as especially problematic. These selected target groups include the so called "Second Generation" or foreign nationals born in Germany, a category perceived as a challenge to planning the social infrastructure such as schooling and, increasingly, as a potential threat.[76] This group became the main target group for the emerging field of foreigner education. Among the "Guest Worker" nationalities, a process of internal diversification produced a racialized hierarchy of foreigners where Turkish (and to a lesser degree Yugoslavian) nationals became more and more singled out as the main problematic group; among these, women and young men became target groups of scientific knowledge production and political intervention. A third risk group emerged from the asylum complex which produced knowledge around the newly created status of "asylum applicant."[77] Finally, from 1983 onwards, return migration and related research

74 Heckmann 2013, 37ff.

75 Eichenhofer 2013, 45 f.

76 Wilpert 1984, p. 310

77 Bade 2016, 90f.

became another core element of migration policy-making and knowledge production.

## Legitimatory Knowledge

When analyzing the symbolic level of knowledge production during the "Lost Decade," a shift away from technocratic, labor-market orientated arguments can be discerned. By the end of the 1970s, a well pronounced problem-perspective on migration and related social phenomena emerged which can be regarded as typical for the discipline until today. The 1981 representative survey lists several demographic factors such as a growing foreign population, especially children, as the main reason for "unfortunately not reduced, but rather increased and novel social problems of foreigners."[78] This focus on problems of integration is clearly visible:

> "Much research started from the more or less unquestioned assumption that labor migrants and their families cause problems and are confronted with a number of social problems due in large part to their inadequate capacity to integrate. In other words, the immigration and settlement process of labor migrants and their families were not conceptualized as an internal and foreseeable permanent socio-structural element of society but rather as an unintended external element affecting 'German society' which needed to be adapted to the existing structures."[79]

The problematization can be seen as the cognitive outcome of the notion of "Germany is not a country of immigration," since it conceptualizes the presence of a sedentary foreign population as a temporary and exogenous problem of societal integration. This problem perspective is most clearly formulated in pedagogic research of the 1970s called "foreigner education." As applied research, it focused on the rectification of foreigners' deficits in schooling contexts.[80] One of the most influential concepts of this research is the idea of "national classes" which is based on understanding ethnic diversity as a deficit to be overcome by reducing the heterogeneity within national classes on the one hand and eradicating the deficits of the foreigners – especially poor German language skills – on the other to prepare them for schooling in

78 Der Bundesminister für Arbeit und Sozialordnung 1981, II

79 Bommes 2009, p. 139

80 See Mecheril et al. 2010, 16ff. for an overview

"normal classes." More simply put, in addition to the perception of ethnic plurality as a deficit, knowledge production in foreigner education is subject to a process of thorough "pedagogization", as Griese notes in 1984. This process distributes social roles to Germans and foreigners, therefore reinforcing the divide between the two groups. Germans emerge as pedagogical problem-solvers, whereas problems are thoroughly explained as foreigners' deficits. The reproduction of societal problems is thus conceptualized as unintended consequences of principally benevolent, engaged educators:

> "A new profession and discipline has been (successfully?) established, gains increasing influence and attention and is about to eliminate societal dysfunction (the so called 'foreigner problem', the 'time bomb', the 'social explosive') by political mandate without realizing that it contributes to the production of these problems."[81]

Indeed, some of the methods of foreigner education consistently failed to reach the stated aim: enhance the schooling success of foreign-born children. On the contrary, children attending "national classes" were even less likely to leave school with a diploma than those who had been attending regular classes, therefore provoking the critical notion of a "two-tier educational system." At the same time, social research produced relatively consistently an internal diversification of the foreigner population, according to which Italian, Spanish, Portuguese and Greek national groups gradually normalized in their socioeconomic features, while Yugoslavian and especially Turkish nationals displayed a comparatively worse performance across almost all socioeconomic indicators: statistics on schooling, income, job position, housing situation, and language acquisition displayed a growing diversification between the two groups of nationals. In the conceptual and theoretical models of foreigner education, with a focus on institutional development and the elimination of ethnic diversity through homogenizing schooling methods, these increasing differences were hard to explain. As a result, the frame of reference for knowledge production changed by about the beginning of the 1980s by increasingly focusing on the individual behavior of migrants. In its initial impetus, this turn has been regarded by contemporary sources as progressive: migrant action was for the first time part of the academic discussion, which was believed to correct patronizing images of foreigners as inherited

81 Griese 1984b, p. 5

from "Guest Worker" research. The "individual turn" shifted the focus of attention away from an institutional perspective which conceptualized migrants as rather passive recipients of educational programs directed almost exclusively at correcting their deficits.

In reaction to this, the newly-introduced perspective of difference promised initially to overcome the narrow focus on social problems and the implicit conceptualization of foreigners as inherently problematic; in this context, differences between national groups were explained usually as a result of socio-economic circumstances, especially selective legal discrimination.[82] Hartmut Esser's oeuvre can be regarded paradigmatic for this shift of knowledge production: in 1980, for his habilitation, Esser proposed an approach of "methodological individualism" which explained integration not as a series of generational transformations, but as a result of individual investment choices.[83] Differences between national groups were explained by differences in human and social capital, and especially by differences in the legal framework of chances and restrictions. However, this approach changed rather quickly to a culturalist style of argumentation: "after 1983, the legal barriers which had been central to argumentation disappear from the scene."[84] The individualistic turn and especially the increased focus on the Turkish sub-group emphasized personal features of migrants over structural factors and presented a different explanation for structural inequalities between national groups in the concept of culture.[85] Cultural identity, understood in an essentialist way, was thought to heavily influence the way of life, mental structure, and correspondingly the schooling and labor market success of migrants.[86] In this context, research reports construct cultural difference as a problem for policy-making: "[...] what seems morally unacceptable to foreigners, or emotional or hot-blooded to Germans is an expression of cultural difference. The understanding of this difference can only be achieved if the distance between the groups is lowered."[87]

While on the surface, the tone of argumentation is less paternalistic and pathologizing as compared to "Guest Worker" research, the argument never-

82 Hetfleisch 2017, p. 94, Bommes 2009, p. 141

83 Wimmer 2009, 319ff.

84 Hetfleisch 2017, p. 94

85 Lanz 2007, p. 82

86 Ibid., p. 85

87 Der Bundesminister für Arbeit und Sozialordnung 1981, 518f.

theless supports the conceptualization of (both the German and the foreign) culture as a stable and immutable personal feature which was furthermore fairly homogenous within a given national group.

Culture was thought to be mainly transmitted by the family and to some degree by educational institutions, which is why culture-based arguments emerge most prominently in these two contexts. In foreigner education, the grim fate of the "Second Generation" was discussed in this way: while younger migrants could in principle be assimilated by schools if immigrated early enough, adolescents were, according to this concept, in the worst situation because they were thought to be torn between two cultures and therefore marginalized two-fold, dis-integrated both in Germany and in the country of origin upon eventual return. In the context of family, the culturalization of knowledge on migrants becomes visible in the discussion of what is regarded as "traditional role models" of men and women especially in the case of Turkish migrants:

> "The high share of married workers can be explained with the situation of women in Mediterranean countries. There, strong ties exist between the family and women and girls; a married woman leaves the family area only in rare, exceptional cases."[88]

In the quote, the foreignness of culture serves as an explanation for a rather common phenomenon – a high share of married women in a young age group. Both the recruitment policies (which created more job opportunities for men than women) as well as legal requirements of marriage for family reunification are left unregarded as an explanation for high marriage rates among Turkish migrants. Instead, ancient cultural patterns are presented as an explanation, a reasoning which is curiously absent from earlier reports of the 1950s and 1960s. This reasoning represents a trend of culturalization which formulates questions of family life and gender as an expression of a stable, ex-ante defined culture. One notable result of the "culturalist turn" is an increasing focus on Turkish women. Increasingly, these persons constitute a separate target group for social research because of the assumption that women, struggling with a paternalistic culture and participating increasingly less in the labor market, would be especially vulnerable to social marginalization.[89] Again, the discussion of culture obscured the tightened

88 Ibid., p. 67, own translation

89 Angenendt 1992, 178ff.

legal prescriptions which can at least partly explain the phenomenon under scrutiny: since 1973, family reunification was the only possibility to immigrate. The same is true for social isolation and the retreat from the labor market, which can likewise be well explained by restrictions in the access to the labor market to dependent family members.

To explain differences in socio-economic integration, a hierarchy between foreigners emerged, based on a concept of "cultural distance," which was ultimately based on a perception of otherness. Former "Guest Workers" from European states such as Spain, Italy, or Greece – only 25 years earlier paradigmatic representatives of the foreigner – were re-conceptualized as culturally less distant and therefore less foreign; Turkish and Yugoslavian nationals, together with Asian and African asylum seekers, were characterized in contrast to this as "real" foreigners who were to a lesser degree able to integrate into society.[90] The "cultural distance" model explained both the differences between national groups and justified to a degree the ongoing and increasing selective legal discrimination especially towards the latter group. While differences in legal statuses were normally part of reports on foreigners, they did not usually serve as an explanation for socio-economic differences; instead, the cause-and-effect relationship was often turned around and the lack of more stable legal titles were explained by the lack of language skills, or knowledge about administrative processes on the side of the migrant, respectively. As an effect, the quite selective application of legal discrimination measures was therefore to a large degree invisibilized from governmental knowledge production on migration through the framework of cultural difference between national groups. Furthermore, the argument of integration problems of older children was one prominent reason to restrict family reunification of these children; cultural distance could therefore be successfully employed to justify restrictive policy measures in the name of integration.

## Conclusion

In sum, governmental knowledge production during the "Lost Decade" is characterized by an increasing diversity, which reflects on the one hand the diversification of migration streams and countries of origin, but also on the other hand a growing internal diversification of the foreigner population due to unequally distributed legal, economic and societal resources. The

90 Lanz 2007, p. 82

first factor had a considerable impact on the structure of the knowledge created in the sense that a larger framework of reference is largely absent. Instead, selected target groups which are perceived as especially problematic are identified and singled out in terms of legislative action and knowledge production, while other non-nationals are made "invisible" in research and policy-making, most importantly EC-nationals.

Regarding the production of symbolic knowledge, a distinct trend from macro-economic, labor-market oriented arguments to culturalistic, demography-based arguments is visible throughout the "Lost Decade."

## Instrumental Narratives and Institutional Traditions

The history of governmental knowledge production on migration is illustrative in two respects: on the one hand, the BAMF's self-understanding as a successor to earlier administrative and conceptual endeavors becomes apparent. On the other hand, the processes of creation and development of key concepts and terms which today constitute the intellectual foundation of migration research have been described. In respect to the BAMF's self-understanding, it becomes clear that this history is read through the lens of an instrumental approach to knowledge utilization: in all phases of research, technical information to policy-making is stressed, whereas legitimatory and especially politically controversial knowledge is omitted or downplayed.

If the phases of knowledge production are compared to each other, certain differences – but also similarities – can be discerned. The BAMF's construction of the three phases of research follows at first glance a relatively stringent logic of relevance through volume: the migration streams in question are most important in terms of numbers from the respective eras, hence the governmental interest in creating knowledge about them. However, it became clear that the BAMF's self-image is only a partial reflection on the historic events; most importantly, the idea that resettler research, foreigner research, and integration research represent historic successors of the study of essentially the same social phenomenon is a product of relatively recent historical analysis. It creates an ex-post order of knowledge which links policy fields which had hitherto not been understood as different forms of the essentially same social phenomenon.

Also, analysis of the governmental knowledge shows that it has in fact more than the instrumental function for administration as suggested; instead, for every phase, strategies of legitimization, problematization and suggestions for future policies point to a symbolic function of knowledge which transcends immediate practical application in policy making. While this finding suggests that it is doubtful that knowledge production follows the instrumentalist ideal as proposed by the BAMF, this does not imply that knowledge production per se is irrelevant for policy making. In the contrary, the fact that governmental knowledge has been produced across all historical eras under scrutiny speaks for its relevance. In the analysis, some alternative uses of knowledge have been outlined, most importantly through its discursive functions such as calming the public debate or legitimizing policy. These findings will be further utilized in the analysis of contemporary knowledge production in the remainder of this text.

When comparing the phases of governmental migration research in respect to the formal and institutional set-up, it seems clear that policy measures were much less accompanied by formal research in the case of foreigner research than in the case of resettlers and expellees. While during the 1950s, the expected social and economic difficulties of integration led to the construction of a rather elaborate statistical and scientific surveillance apparatus, the efforts in connection with the "Guest Worker" recruitment system have a much lower profile. After all, the target population was quite small initially: "Guest Worker" migration was discussed within the framework of foreign labor in general and gained only by the mid-1960s higher significance. In contrast to that, the large volume of resettler immigration was a known fact from the outset and the prime reason for installing an elaborate statistical and scientific monitoring system to begin with. The annual reports by the Federal Agency for Labor and statistical data published by the Federal Statistical Office since 1967 were the only relevant sources of governmental knowledge on "Guest Workers" at the time.[91] The reports of the Federal Agency show quite detailed reporting of statistical data not unlike comparable reports on resettler integration. Unlike the latter, however, the perspective is strictly confined to labor and employment. This narrow focus could not account for the growing importance of social processes outside of labor market contexts such as the diminishing administrative control of the job placement system. All in all, the amount and quality of governmental knowledge produced and published by the government about migration is quite limited during the 1960s and 1970s. This fits to the low institutional profile of migration policy-making which is

91 Statistisches Bundesamt 2012, p. 4

organized primarily on the level of federal administration, not as a ministry like in the case of resettlers, and outside of parliamentary control.[92] While the topic of migration gradually became more politicized during the "Lost Decade," the low institutional profile of governmental knowledge production remained relatively constant.[93]

When analyzing the symbolic knowledge generated in the different phases, some remarkable differences emerge. During "Guest Worker" recruitment, the impersonal, somewhat technocratic style of policy legitimization is a quite striking difference to resettler and refugee research. This is because the latter is deeply rooted in a tradition of ethno-national historic research which evoked the "national community of fate" as a legitimization for the redistribution of resources to the various refugee/expellee status groups.[94] Consequentially, there have never been macro-economic examinations if, for example, the resettler support paid off economically. In contrast, in the "Guest Worker" phase, arguments are constructed around abstract objects like "labor force," "shortage of labor," "economic upswing" and "depression," or similar concepts from macroeconomics. It seems that no longer "ethnic belonging" (*Volkszugehörigkeit*) but rather macroeconomic principles (*Volkswirtschaft*) legitimize the policy in principle. The understanding of policy-making as planning can be understood as a rather paradigmatic expression of contemporary political theory.[95] This principle, together with the rather unprecedented power of the state administration in policy making, can be considered the two main characteristics of the "Guest Worker" system and at the same time the two most important differences to resettler migration. This discursive structure changed again at the beginning of the 1980s: through analytical tools of demography, the focus of knowledge production shifted from the economy to the foreign population in the demographic understanding of the term. Together with the assumption of the central responsibility for foreigner's policy by the Ministry of the Interior and a corresponding trend of securization, knowledge production increasingly

---

92 Schönwälder 1999

93 For example studies in the area of foreigner education (Cp. Griese 1984a) or government-sponsored research in the context of the return migration support policy (Cp. Hönekopp 1987a).

94 Bommes 2009, p. 129

95 Schneider 2010, p. 40f.

focused on narrowly defined target groups perceived as especially problematic through a lens of cultural difference: these include especially second generation youth, Turkish women, and non-European asylum seekers. All in all, through selective processes of knowledge production, social integration, and legal differentiation, a racialized hierarchy of foreigners emerged in the course of the "Lost Decade". While during "Guest Worker" recruitment, "Guest Workers" were referred to collectively as a social group, this unified category was thus broken up: some of the former foreigners were no longer perceived as foreign and enjoyed economic and social rights comparable to German citizens, while others were perceived as even more foreign than before, especially visible minorities.[96]

The governmental character of this knowledge is visible in two key mechanisms: first, in the creation of terminology and second, in the internalization of political standards to research. The first point refers to the fact that most research is based on governmental statistical concepts, either as sources for quantitative data or in relation to the population groups it analyzes. By using legal definitions, the many possible interpretations of what could, for example, constitute a refugee or a "Guest Worker" become a particular, hegemonic interpretation – that of the state. Methodically, this means that definitions and data sources are extracted from legal norms and administrative statistics; research questions are usually formulated from the perspective of the administration, and are furthermore framed by the available data. In scientific research as well as government statistics, the main unit of the population in question (foreigners) and the main unit of comparison (nationality groups) remained the standard method of measurement since the mid-1960s, although it became to a degree less meaningful through socio-economic, ethnic and legal diversification. Different status groups, stemming from different migration processes such as asylum seekers, second generation descendants of migrants, former "Guest Workers" and their spouses were grouped together in one national group without further differentiation according to their legal status. Epistemically, this practice became especially problematic in cases where different migration streams originated from the same country, such as refugees from Turkey in the aftermath of the 1980 military coup, or civil war refugees from Yugoslavia in the 1990s. Furthermore, the increased diversification of countries of origin produced a gradually enlarged and diverse "other" group in the foreigner statistics and governmental reports, corresponding to

96 Lanz 2007, p. 82

a general decrease of the share of the four large "Guest Worker" nationalities in the total foreign population.

The second point, internalization of political standards, stems from the structural conditions of knowledge production in migration research. Until the end of the 1970s, a joint perspective of researchers and government towards the object of policy and research is discernible: both resettler and foreigner research are examples of applied research which conceptualizes the research field through a legal-political lens. The joint perspective on the research subject is expressed in the notion of "integration" which is maybe the most stable common thread throughout the phases of research.[97] Integration is understood as a pre-given desideratum both in political and in scientific terms; the "dual nature" of the term as a normative value and a scientific concept reinforces the theoretical foundations of the term and lends it "intuitive plausibility."[98] This means that, for example, the question of a correct distribution key for resettlers, or the cost-benefit equation of "Guest Worker" recruitment, is usually formulated from a governmental point of view. The scientifically correct solution to a given problem argues from the perspective of "the" economy or "the" society.

The conceptualization can in some cases be identified as a source of blind spots in the knowledge: for example, the structural reduction of personal features of "Guest Workers" to the question if they are beneficial for employment or not inhibit the analysis of long-term trends such as the emergence of a migration network and lead to inconclusive policies as in the case of family reunification. Another negative effect can be described with the critique of methodological nationalism: both resettlers and "Guest Workers" are constructed as a national particularity unfit for international comparison or even theoretical reference.[99] For example, "Guest Workers" are not conceptualized in the context of European work migration schemes which were implemented across most western European countries in the after-war period, but are instead seen as the sole outcome of bilateral recruitment contracts and therefore by definition incomparable to other migration phenomena, especially to so-called "classic immigration countries."[100] Likewise, resettlers are conceptualized as a result of a unique German history in Middle and Eastern Germany,

97 Cp. Bommes 2009, 162ff.

98 Bommes 2012, 19f.

99 Heckmann 2013, p. 36

100 Castles 2000, 29f

as well as the post-war order, which therefore inhibits theoretical references both to "Guest Workers" and to other migration processes. However, similar recruitment (and resettlement) schemes were developed in most European countries.[101]

The functional distribution of roles in this model is separated between knowledge production and political action: while the "primary role of sociologists is to study, chart, and offer remedies to social inequality,"[102] government was expected to act upon these problems once the solution was identified. All in all, the common perspective of the researcher and the scientist in conceptualizing a research/policy object can be seen as a core feature of governmental knowledge production. Together with an instrumental understanding of science – as a welcome and necessary source of information to the policy maker – this understanding of knowledge lies at the foundation of the BAMF historiography.

101 Scholten 2011a, p. 80

102 Favell 2001, p. 360

# Structural Conditions of Knowledge Production

The political reform debate and ensuing policy changes between the change of government in 1998 and the passing of the Residence Act in 2004 are often characterized as a "paradigm change" in migration policy-making in government documents and historical migration research. During this period of political change, the Research Group was founded, a process which will be analyzed in the following chapter. For the analysis, two findings in the last chapter are of particular importance. First, as is evident from the BAMF's historiography, there is a strong reference to instrumental theories of knowledge production. This narrative draws a picture of rational, objective research as impartial information to policy-making especially in technical matters, while relegating the production of symbolic knowledge to the political arena. While it could be demonstrated that this discourse is in several respects imprecise – technical knowledge is political, and policy guidelines are influenced by technical knowledge – this narrative can be considered extremely influential for the creation of the Research Group. Second, the analysis of the history of governmental research revealed different styles of knowledge production and governance which continue to influence knowledge production until today. The mechanisms influencing knowledge production stemming from institutional, political, and scientific factors have been analyzed from the different phases of the history of migration research. In a similar manner, institutional, political, and knowledge-related factors will serve as a basis for the analysis of the Research Group's scientific output. In both these respects, the tradition of governmental research as laid out by the BAMF can be understood as a structural condition of knowledge production: the methods, data sources, research topics, and political uses of knowledge production constitute a point of reference for how the BAMF Research Group understands its current role.

To complete the overview of structural conditions and practical constraints of governmental knowledge production, the development of the institutional organization of knowledge production has to be accounted for as well. In the case of the BAMF, this development is interesting for two reasons. First, in its self-understanding, the Research Group at the Federal Office for Migration and Refugees, or as it was called until 2003 the "Federal Office for the Recognition of Foreign Refugees," represents like no other government authority the institutional changes of the new "policy paradigm" in migration policy making.[1] However, the new role of the Federal Office was by no means predetermined and without an alternative especially in regard to the establishment of a research body for governmental knowledge production on migration. Second, this history of foundation is discussed in terms of its impact on the formation of structural features of knowledge production at the BAMF. These features are not only influenced by the discursive framework of reference of governmental knowledge, but additionally by the specific institutional make-up of the Research Group and its position within the institutional hierarchy. Especially the latter point has been repeatedly raised during expert interviews when explain research designs or interpretations of findings. The configuration of the Research Group can be traced back to decisions taken in the phase of institutional and political reconfiguration of migration policy-making in the early 2000s. The analysis of the history of the foundation of the Research Group therefore focuses on the question of which role and function the Research Group fulfills at the BAMF and how this configuration frames the production of knowledge.

The chapter is divided into three parts. The first part recapitulates the institutional development of the Federal Office for Migration and Refugees especially in regard to its enlarged competencies in knowledge production in the wake of the institutional reconfiguration of the early 2000s. This reconfigured structure will be more closely examined in the second part of the chapter by describing the development of a specific self-understanding of the Research Group. This *Selbstverständnis* ("self-understanding") nicely illustrates both structural-institutional conditions as well as conflicts of interest between various institutional actors, most importantly the BAMF leadership and the Ministry of the Interior, over the question of the strategic orientation of research.

Sources for this analysis are, besides expert interviews, BAMF publications and documents. Especially the Research Group's yearly reports provide some insight into the development of the formal structure of the research

---

1 Kreienbrink and Worbs 2015, p. 330

unit; furthermore, the development of a mission statement can be retraced with the help of these reports.

## Paradigm Change

In 1998, the newly elected red-green government attempted to replace the "no country of immigration" paradigm. In this context, the "Lost Decade" was portrayed as a negative example in terms of inapt knowledge production: according to the BAMF's historiography, the political stalemate was produced by the failure to realize the permanence of foreigner settlement migration in Germany. In abstract terms, the period is characterized by a growing antagonism between "objective" science and "irrational" politics: "The changed situation has been reflected by foreigner research and called for a redefinition of migration policy. Politics, however, does not acknowledge these new developments, and invents the formula 'Germany is not a country of immigration'", which is maintained until 1998."[2]

This context is important for the ensuing reform period: around the turn of the millennium, the notion of paradigm change "was in the air."[3] This paradigm change was most importantly connected to the hope that political stalemate could be overcome with expert knowledge and scientific policy counseling – with other words, a classic instrumentalist narrative was drawn.

The paradigm change is usually connected to the work of the Independent Commission "Integration" and the commission's final report in 2001.[4] One reason for this lies in the political constellation and the principles of operation of the commission: the body consisted of experts on migration from academia, politics, and civil society. It was led by conservative politician Rita Süßmuth, a decision intended to soften the expected resistance of her own party. To fulfill the promise of independent expertise, the commission's work was obliged to scientific rather than political standards of quality. In fact, only a small minority of the experts participating in the hearing process were politicians, while most of the experts were either scientists or mid-level administrative staff from government authorities.[5]

2 Heckmann 2013, 38f.
3 Interview, December 2017
4 Unabhängige Kommission "Zuwanderung" 2001
5 Numbers quoted after Schneider 2010, p. 258

Furthermore, 18 studies with an impressive total of more than 2,000 pages were commissioned.[6] All in all, the working mode of the commission promised scientifically-grounded policy proposals, published in a report, as a basis to decision-making. This mode of knowledge production by itself can be interpreted as a demonstrative turn from the "no country of immigration" dogma.[7]

Conceptually, the notion of expert knowledge retained a central position in the Independent Commission's proposals for policy reform. Indeed, independent expert knowledge emerges as the main remedy to the pathologies of migration policy-making of the past. Consequently, many of the proposals are based on the principle of scientific expertise on all levels of policy making. The proposals aimed at transforming not only the institutional structure, but above all the style of policy making: not ideological controversy and administrative *muddling through*,[8] but rather independent expert knowledge should become the chief governing principle of migration policy making.

In terms of policy reform, the commission proposed a policy of planned immigration. This included most importantly a coherent system of immigration steering and control similar to the Canadian model of a "point system."[9]

This institutional structure was largely based on an expert opinion by Klaus Bade.[10] According to this proposal, administrative responsibility for all matters concerning migration and integration were to be concentrated in a single authority, the Federal Office for Migration and Refugees. Expert knowledge was to become a systematic part of policy-making, including among other things the introduction of a comprehensive statistical information apparatus, the coordination of university research activities, the evaluation of legal acts and other policy items, and the prognosis of future migration movements. Institutionally, this research was to be organized in a two-tier structure consisting of the *Zuwanderungsrat*, an independent council of scientific experts, as well as the *Bundesforschungsinstitut für Migration und Integration*, a departmental research institute.[11] The Immigration Council's most important task was the preparation of a yearly report with immigration quota based

6 Schneider 2010, 253ff.

7 Ibid., p. 364

8 Lindblom 1959

9 Unabhängige Kommission "Zuwanderung" 2001, p. 84

10 Schneider 2010, p. 260

11 Unabhängige Kommission "Zuwanderung" 2001, p. 286

mostly on labor demand. The proposal followed the structure of resort research institutions in socio-political administrative areas, most importantly the *Bundesinstitut für Bevölkerungsforschung* ("Federal Institute for Population Research") and the *Institut für Arbeitsmarkt- und Berufsforschung* ("Institute for Employment Research"). Both institutions are associated with federal authorities (the Federal Statistical Office and the Federal Agency for Labor, respectively) and provide scientific research for direct application in policy-making.

## Implementation

The Independent Commission proposed to turn the BAMF into one of the central administrative authorities in the field of migration policy, committed to the ideal of knowledge-informed policy-making. The central legislative measure implementing this reform was the Residence Act designed to contain most of the Commission's reform proposals. However, despite the government's political backing and the Independent Commission's effort to propose impartial and scientifically grounded reform concepts, the process of legislative implementation turned out to be rather bumpy. Some parts of the legal act were declared unconstitutional by the Supreme Court. During the implementation process, the government lost its majority in one of the legislative houses (*Bundesrat*) which caused a political deadlock. In the meantime, parts of the reform proposals were implemented by way of executive decree, among other things the Immigration Council in April, 2003. The council began by first using its expertise during the political controversy over the Residence Act. The Immigration Council's report was strongly focused on the results of the Independent Commission's work: it underlined the importance of scientifically grounded policy-making in general and criticized the lack of political vigor to implement knowledge-based policy-making principles.[12] Moreover, the report proposed moderate immigration according to the needs of the labor market. However, especially the proposition of immigration sparked fierce media criticism. As a result, the council was finally dissolved due to heavy resistance of the conservative opposition parties.[13] After the removal of the Immigration Council, political compromise was established and the Residence Act was finally passed by both houses of the parliament.

---

12 Sachverständigenrat für Zuwanderung und Integration 2004, p. 395

13 Interview, December 2017. The council was finally dissolved in December 2005.

As a result of the legal reform, this is the actually implemented institutional configuration of governmental knowledge production: the pre-reform hierarchical order of institutions remains largely intact, the BAMF remains a subordinate authority to the Ministry of the Interior; the Ministry of the Interior keeps its central coordinating role in policy-making on most matters in the policy field. Nevertheless, the Federal Office for Migration and Refugees constitutes something like the epicenter of the reforms: Not only does it retain the important competencies in integration policy, but also the newly established research unit. The Residence Act contains a legal mandate to produce scientific knowledge at the BAMF (Section 75 Residence Act):

> "[…] The Federal Office for Migration and Refugees shall have the following duties: […]
>
> (4) conducting scientific research on migration issues (accompanying research) with the aim of obtaining analytical conclusions for use in controlling immigration."[14]

Quite remarkably, the Research Group – as a dependent unit in the BAMF hierarchy, not as a separate institution – is by and large the only element of scientific knowledge production which has actually been implemented in law. While this aspect will be discussed in some depth later on, it is important to highlight some institutional features of the Federal Office for Migration and Refugees.

Judging from the history of the Federal Office for Migration and Refugees, its future central role in governmental knowledge production after the institutional reconfiguration of the "paradigm change" was relatively surprising. This central position is quite the contrary to the rather marginal role it had always played in policy-making and knowledge production on migration. The authority was established in 1953 as a successor of a government bureau for the material support of Displaced Persons.[15] After the passing of the Foreigner Act in 1965, the authority was renamed into *Bundesamt für die Anerkennung ausländischer Flüchtlinge* ("Federal Office for the Recognition of Foreign Refugees"). The office's tasks consisted chiefly in the processing of asylum claims. At the time of its foundation, asylum migration was small in volume compared to other migration streams such as expellee, "Guest Worker" recruitment, or ethnic German migrations from the GDR and Eastern Europe. Until the end of the

---

14 Bundesministerium des Innern, Section 75

15 See Kreienbrink 2013 for an overview of the history of the Federal Office

1970s, the number of asylum decisions rarely exceeded 10,000 per year, with occasional peaks in the wake of political upheavals. As a result, the Federal Office's area of responsibility was quite confined both in terms of workload and competencies. Although the number of asylum applications slowly increased during the 1980s, the surrounding conditions defining the area of competence for the Federal Office remained stable for a large part. Asylum policy was mainly structured by the political conditions of the Cold War: refugees from Eastern Europe and ethnic German resettlers were generally assumed to be politically persecuted, which resulted in preferential treatment of these refugee groups. This assumption was also useful in a political sense since it was seen as a proof of the superiority of Western liberal democracies.[16] This relatively preferential treatment is contrasted with the creation of the new status of "Asylum Seeker" as described in the last chapter: asylum seekers from Africa, Central Asia, and Turkey were increasingly subject to a racialized, pejorative political discourse and repressive treatment by the authorities. This included the erection of selective hurdles to the access to asylum to limit the influx of asylum seekers especially from Turkey, but also from Afghanistan and Pakistan in the early 1980s. The discourse of "asylum abuse" contributed to a process of curbing asylum inflow with repressive measures in an increasingly critical, political climate surrounding asylum and the work of the Federal Office throughout the 1980s. In sum, the BAFl represented perhaps like no other government the "no country of immigration" dogma of the "Lost Decade".

The precarious stability guaranteed by the political conditions of the Cold War shifted by the end of the 1980s in the wake of the beginning downfall of the Soviet Union and its allies. The consequences of this crisis were expressed in an historic surge in asylum applications and a dramatic increase of the workload and the backlog of asylum decisions at the Federal Office. In the years before and after 1990, up to 450,000 applications of asylum were registered annually. This surge resulted in a multifold increase of personnel at the Federal Office – from several hundred government officials to 5,100 employees in 1993.[17] The increase in personnel was rather difficult to implement at the time: the requirements of asylum deciders usually require fully qualified lawyers. However, the labor reserve for these employees was practically swept

16 Klekowski von Kloppenfels 2003, p. 400

17 Including 1,000 government workers delegated temporarily from other authorities. Cp. Kerpal 2003, p. 12

empty as a result of the massive recruitment of state officials in wake of the German reunification. The resulting shortage of personnel could only be overcome by lowering the requirement standards[18] of recruitment and granting exceptionally generous conditions of employment.[19]

In 1993, the political conditions of the Federal Office's work fundamentally changed again: after the factual abolishment of the constitutionally guaranteed right to asylum, the number of asylum applications sank rapidly. From a peak of almost half a million applications in 1990, the number of newly filed asylum claims dropped in 1993 and hovered around 50,000 annually during the rest of the decade. As a consequence, the Federal Office with its oversized workforce and its network of almost 50 field offices underwent restructuring again. Until the end of the 1990s, personnel was cut by half to 2,500 staff in 22 field offices. The privileges granted to newly hired employees constituted a difficulty in this process and effectuated a somewhat negative selection of personnel; while younger, well qualified employees could be delegated to other state institutions, older, less qualified and less mobile staff stayed at the Federal Office. By the end of the 1990s the Federal Office for the Recognition of Foreign Refugees had clearly experienced a rather untypical phase of institutional restructuring. The declining numbers of asylum procedures, reduced staff, and the relatively small area of responsibility pointed to a gradually declining degree of institutional importance in general. This trend is perhaps best illustrated by the anecdote of a state official who was delegated to the BAMF as a trainee in the mid-2000s by appointment of his superiors. This he experienced as a mild form of harassment due to his party affiliation.[20] In conclusion, until 2005, the office was characterized as a relatively unimportant "institutional backwater,"[21] both in geographical and hierarchical terms remote from the institutional centers of political decision-making.

## Establishment of the Research Group

During the course of the reform, the Research Group came into being at the end of 2004. Research started in two units: migration and integration re-

18 Field Notes, background talk with a former BAMF officer, 2013

19 Kreienbrink 2013, p. 406

20 Field Notes, background talk with a former BAMF officer, 2013

21 Boswell 2009b, 163f.

search were initially conducted in a single unit, while a second unit was concerned with the economic aspects of migration.[22] This structure reflects to a degree the internal organization of the Independent Commission whose general office was located at the BAMF. The two working groups of the Independent Commission drawing most heavily on scientific knowledge, labor market and integration, reemerge as research units in the BAMF. Furthermore, the idea of merging migration and integration research refers to a frequently quoted key phrase of the Commission's report, according to which "integration and migration are two sides of the same medal."[23]

Some researchers were transferred from the Independent Commission's staff office or were recruited among the experts who participated in the commission's hearings.[24] Other researchers were recruited for the task of compiling the Migration Report, a research project regarded as one of the most prestigious at the time.[25] All in all, considering staff and research projects, the Research Group was able to take over a large share of the resources and functions of the now inactive Immigration Council.[26]

The actually implemented institutional set-up did not only fall short of the Commission's proposals in terms of institutional independence, but also created a situation of ambiguity regarding the actual mandate of the Research Group: although the conduction of research was codified in the Residence Act, the wording of the paragraph was rather imprecise. The legal text mentioned "accompanying research" which was to be conducted to "obtain analytical conclusions for use in controlling immigration."[27] At first glance, this expression seems to fit the concept of instrumental research in accordance with the proposals of the Independent Commission; indeed, most of the wording is taken over from the relevant parts of the Commission's report which likewise mentions "accompanying research" as one future task of migration policy making.[28] However, unlike the Independent Commission's paper, the term "accompanying research" is not linked to the established concept of departmental research (or to any other concept of knowledge production, for that matter)

22 Research Notes, Email from the Research Group, February 2014
23 Bade 2001, 18f.
24 For example, Hans Dietrich von Loeffelholz. Cp. also Kreienbrink and Worbs 2015, p. 326
25 Boswell 2009b, p. 180
26 Ibid., 167f.
27 Residence Act, Section 75
28 Unabhängige Kommission "Zuwanderung" 2001, p. 292

and is not further specified as such. In effect, the concrete meaning of the research task and therefore the functions of the Research Group were quite unclear.

In this situation, the Research Group drew up the initial research agenda mainly according to internal deliberations. Apart from two studies which were already commissioned by other government authorities concerning the Muslim community in Germany and the effects of a law restricting residence rights of Ethnic Germans, most first-hour study topics were selected by the Research Group. The development of this approach to interpreting the legal mandate to accompanying research – drawing up research projects according to the criterion "what could be of interest" – will be discussed in some detail later.

The rather unusual degree of institutional liberty connected to the blurry legal mandate for accompanying research was not, however, experienced as an unequivocal advantage by the first-hour staff. The combination of hierarchical subordination of the research with an unspecified task created a situation where the Research Group was isolated from the rest of the Federal Office not only in functional, but also in cultural terms. In practice, these tensions materialized in several ways, most of them concerning the different work cultures of researchers and government officials:

> "Very important thing, office hours and time stamp cards. The idea of punching the clock was unfamiliar to the researchers at the time. But it was impossible to obtain permission for exemption, because there were worries that researchers would be even less accepted if they had a privileged role. There was an element of incomprehension on the side of the Office: someone only because he went to university comes into the Federal Office and gets fairly well paid [in comparison to government officials, VK]. They write studies which take a long time until they are done and no one knows what they are good for."[29]

29 "Ganz wichtige Geschichte, Kernzeit und Stechuhr. Die Vorstellung, dass wissenschaftliche Mitarbeiter einstechen, war, sämtlichen Wissenschaftlern völlig fremd damals. Sie konnten es aber nicht durchsetzen, dass die Forscher eine Sonderrolle kriegen, weil man befürchtet hat, dass es noch weniger Akzeptanz gibt. [...] Und, eine Komponente spielte da auch mit, ein Unverständnis auf Behördenseite, warum jemand, nur weil er studiert hat [...], jetzt plötzlich in diese Behörde kommt und echt gut bezahlt wird und an Papieren schreibt, die noch dazu sehr lange dauern bis sie fertig sind und wo man nicht weiß, was das alles so bringt."(Interview with a BAMF researcher, 2016)

In the quote, the apparent difference between scientific and administrative work cultures is mentioned as a core problem in the initial phase of establishment. This entails both incomprehension of administrative procedures and norms (such as the time stamp) on the side of the researchers as well as incomprehension of the actual task of the Research Group on the side of government officials.

One important aspect of this miscomprehension was the plan to publish research results. This concept followed from the strategy of imitating practices of departmental research: these institutes publish research results routinely, and similar practices prevail in ministries and other government authorities with commissioned scientific studies.[30] Additionally, the idea was based on deliberations to make the job postings at the Federal Office more attractive to scientists by offering the possibility to publish to a scientific audience. To achieve this, a series of Working Papers was established.[31] However, considering the usual practice at the BAMF, publication of research results was a stark deviation from the Office's common treatment of public relations. This traditional approach is characterized by a rather unusual degree of restraint in terms of public visibility:

> "Most officials working on asylum or integration within the Federal Office and the Interior Ministry attach little weight to external output. As a senior researcher explained, when the Research Group was first established it was not clear to many Federal Office staff that there should be external publications at all. 'Some people thought it was sufficient to send reports to the Interior Ministry.'"[32]

This practice of avoiding public attention is understandable given the history of the policy field in combination with the subordinate position of the BAMF: asylum and migration policy was, at least since the 1980s, a field of increased public attention which often resulted in criticism of the Office's work in the media. This critique was not, however, connected to the BAMF's decisions on policy, since the Federal Office as a subordinate authority had little actual influence on policy-making. Rather, the criticism can be attributed to the bureaucratic proverbial wisdom that "garbage rolls downhill," meaning

30 Kraler and Perchinig 2017, p. 63

31 Boswell 2009b, p. 185

32 Ibid., p. 186

policy failures are often blamed on executive authorities instead of the decision-makers. Given the long-standing history of negative media attention confronting the Office, the general strategy of keeping a low public profile seems reasonable. Although the Federal Office managed its public relations even before 2005, these relations were traditionally treated with some reservation. As a government official put it rather drastically, "media attention is considered an operational accident."[33] The Research Group's strategy of public visibility was therefore one decisive structural difference between the researchers and the rest of the Office which contributed to the initial situation of alienation.

This image of alienation is consistent with empirical research at the Federal Office conducted in 2008 by Christina Boswell. One core piece of evidence in this context is the study on Ethnic Germans. It was one of the first research commissions assigned to the Research Group by the Ministry of the Interior. During the course of one and a half years, a thorough empirical study with representative questionnaires was conducted. However, the study was conducted out of a misunderstanding of the research task, as became apparent later:

> "The two-hundred-page final report was courtly received, but the authors were informed by the Ministry of the Interior that a shorter study would have sufficed and the existing one did not contain the answers to the questions posed by the ministry."[34]

As a consequence, Boswell describes alienation as a structural cultural element of institutions of governmental knowledge production. The case of the Ethnic German study, but also the other observations of diverging traditions of work culture as described above, are interpreted as an outcome of this alienation process. Drawing on Boswell's work, Kraler and Perchinig conclude that the Research Group has failed in their task to bridge the gap between politics and social research.[35] In the literature, in general terms, this failure is often characterized as an expression of a systematic gap between research and politics. Alienation is explained from a system-theory point of view as an incompatibility of inherently different system logics between science and

33 Field notes, October 2013

34 Kraler and Perchinig 2017, p. 75

35 Ibid., p. 85

politics. This reasoning is even part of the official self-understanding of departmental research: according to a strategy paper published by the Federal Government, this type of knowledge production is characterized by "several areas of tension which is caused by different rationalities in science and politics."[36] This reasoning refers to a well-established discourse on structural difficulties of research in bureaucratic contexts; Luhmann's theory of incompatible systems (as in the quote above) is a standard reference in this context. Bourdieu's study on the French national agricultural research institute (INRA) likewise concluded that severe contradictions can arise in an institution which is committed to both knowledge production according to scientific standards and the exercise of political power.[37] From this perspective, alienation between the Research Group and the rest of the Federal Office is an expression of structural incompatibilities which potentially cannot be resolved.

However, the development of the Research Group points to a different, somewhat less pessimistic interpretation: while most interviewees agree that there was initially quite severe alienation, it seems equally common-sensical among interviewees that this gap was to a large degree limited to the first years after the Research Group's establishment:

> "By now, the Research Center has become a normal part of the Federal Office, as a result of a certain process over the years. Everyone has their task and duty, everyone respects each other."[38]

This development is not easy to explain with the gap thesis of governmental research. In interviews, researchers describe a process over some years during which both the Research Group and the rest of the Federal Office gradually adjusted their functions and work cultures. On the side of the Research Group, crucial to this development is the acquisition of processural knowledge about the Federal Office, especially by experienced staff who were transferred from the Immigration Council's office.[39] With the help of this knowledge, the Research Group was in a better position to navigate through the bureaucratic

36 Bundesregierung 2007, p. 3

37 Quoted after Barlösius 2008, 11 f.

38 "Mittlerweile ist das Forschungszentrum zu einem ganz normalen Teil des Amtes geworden, ist ein gewisser Prozess über die Jahre gewesen. […] Alle haben ihre verschiedenen Aufgaben und man respektiert sich." (Interview with a BAMF researcher, September 2015)

39 Field notes, July 2016

workings of the Federal Office and therefore able to carve out established working modes of knowledge production. In this context, the practice of acquiring study commissions, which will be discussed later in some detail, contributed to a functional integration of the Research Group into the BAMF since many of these commissions originated from other BAMF departments, above all the integration unit.[40]

This process of establishment was, however, not only conditioned by the integration efforts of the Research Group's members, but perhaps even more importantly by the structural changes in the Federal Office:

> "The Office has changed extremely in the last ten years. Researchers used to be on the one side, jurists on the other, but this is not true anymore. Especially in the operative area of integration, but also in the asylum department and in other units, there are much less jurists, and more social scientists have been hired. Also specialists on Islam, demographists, geographs, politologists. There has been a radical change in the whole Office."[41]

This "radical change" was also brought forward by the fact that not only the Research Group, but also other administrative departments were founded from scratch in 2005.[42] This introduced new responsibilities, policy tools, and a greater diversity of personnel in terms of professional and academic background and working tasks. The restructuring of the BAMF can therefore be considered a rather exceptional time which serves as an explanation for initial friction between the Research Group and the rest of the Office:

> "If a group of young social scientists is introduced into an institution which is in a process of thorough rebuilding – I am talking about the years 2004, 2005 – there is of course a certain feeling of alienation from the established parts of the Office."[43]

---

40 Gütlhuber and Schimany 2013, 25f.

41 "Das Amt in den letzten zehn Jahren hat sich extrem gewandelt. Am Anfang gab [es die] Juristen und es gab halt die Forscher, aber das stimmt so inzwischen überhaupt nicht mehr. […] Gerade in dem […] operativen Bereich der Integration, aber auch im Asylverfahren […] und […] in anderen Bereichen, sind sehr viel weniger Juristen und es sind Sozialwissenschaftler dazugekommen. Und Islamwissenschaftler, Demographen, Geographen, Politologen. Also, es hat auch im ganzen Amt ein Riesen-Umbruch stattgefunden."(Interview with a BAMF researcher, 2015)

42 This includes most importantly the integration department.

43 "Wenn Sie eine Gruppe von jungen Sozialwissenschaftlern haben, die Sie in ein Amt [einführen], das grade im vollen Umbau ist – ich rede jetzt von den Jahren 2004,5, […]

In this context, it can be assumed that the feeling of alienation as expressed by research staff was not primarily caused by systematic incompatibilities between research and administration, as suggested by the academic literature discussed above. Instead, empirical evidence points to the fact that alienation was caused by the specific context of institutional change which created the development of both the Research Group's tasks and the Federal Office. Researchers were among the first newly hired employees at the office after years of a hiring freeze; they arrived at the Federal Office in a situation where the management was eager to accommodate large numbers of surplus staff with new responsibilities.[44] The restructuring process of the 1990s and 2000s created a staff with relatively old and under qualified members with little perspective on future tasks, a precarious situation where the researchers represented a challenge. The Research Group's undefined task somewhere between departmental research institution and the public relations unit contributed to this feeling of alienation as well, since it did not clarify the question of legal tasks, area of responsibility, and competence which are all highly important for the functioning of any bureaucracy.[45] The resulting dynamics of this process was mainly experienced as a gap in terms of work culture and alienation on the side of the Research Group. However, most interviewed experts confirm that alienation has in the meantime been overcome and that research plays an integral part in the proceedings of the Federal Office.

In sum, the specific constellation of institutional change can be considered the main cause of initial alienation. Members of the Research Group were considered representatives of the institutional change in the Federal Office which challenged established responsibilities and hierarchies. In this context, the Research Group was not in a fundamentally different position than other newly founded departments such as the integration department. This view is further supported by the impression of most interviewees that now most structural differences seem to have somewhat smoothed over. Through the thorough restructuring of the Federal Office, the Research Group repre-

da ist das natürlich gewisser Weise gegenüber Alteingesessenen ein gewisses Fremdheitsgefühl."(Interview with a BAMF researcher, 2015)

44 Boswell 2009b, p. 176

45 Barlösius 2008, 12 f.

sents less of a "foreign body" in terms of the background of its members, its work organization, and its institutional task.[46]

## The Research Group as a Departmental Research Institution

The integration of the Research Group into the Federal Office was described in terms of the structural approximation between BAMF officials and research staff. The process of approximation challenges somewhat the predominant hypothesis of a structural gap between research and policy-making. In a related process, the gradual integration is visible in the development of a self-understanding which can be read as something like a mission statement of the BAMF Research Group. The self-understanding of the Research Group in terms of task, research areas of interest, methods, and aims of research developed over the years. This development is particularly interesting since it demonstrates what kind of knowledge and topics as well as which political uses seem relevant from the researchers' point of view. At the end of the chapter, a specific understanding of political relevance, arising directly from this process, will be outlined.

Again, the blurry legal basis can serve as a point of departure for analysis. Since the legal task of the Research Group was all but clear, initial attempts of self-describing the role and function of government research are drawn up directly from the research work of the group.

> "At the beginning of the year 2005, the Federal Office defined the task of the Research Group. According to this, the aims of the research of the Federal Office are defined as following:

a) Gaining analytic insights for the control of migration
b) Study the effects of migration processes for the Federal Republic of Germany
c) Migration Research in the context of demographic change
d) Analysis of the integration process of Ethnic Germans
e) Evaluation of integration policy measures
f) Study of the economic effects of integration and non-integration."[47]

---

46 Meanwhile, the once infamous punching clocks are abolished in the BAMF. Field notes, December 2016

47 Bundesamt für Migration und Flüchtlinge 2009b, p. 10

In sum, this early mission statement refers almost exclusively to the research agenda and the way it is drawn up. Not incidentally, this list of tasks reads like it had been compiled from the research topics of the first research projects. Apart from the first item on the list which refers to the legal mandate of the Residence Act, all other points simply name research projects in one way or another: "effects of migration," "integration of Ethnic Germans," and "evaluation of integration courses" were each individual study projects.[48] The other two items on the list – demographic change and economic effects – refer to the research areas of the initial two units of the Research Group.[49] Taken together, the list of research goals illustrates the early approach to interpreting the task of "accompanying research" pragmatically from the research tasks; these, in turn, were largely the outcome of a bottom-up process.

However, between 2009 and 2012, a clear trend can be discerned from initially largely self-assigned study topics to an increasing share of mandated research studies commissioned by other state authorities:

> "Initially, we have selected study topics ourselves, under the aspect which topic could be relevant for policy counseling. In the course of the last years, our capacity is increasingly taken up by study assignments. Most assignments originate from the Ministry of the Interior, or directly from the BAMF."[50]

This increasing trend in study assignments is primarily caused by related demands by other ministries, above all the Ministry of the Interior, which developed a rising interest in commissioning research to the BAMF.[51] Over time, it seems that the initial autonomous selection of study topics is completely replaced by commissioned studies; at least in publications, the Research Group seems eager to create this impression. In a 2013 overview of the research activities, the planning of the research agenda is described as a "result of a

48 Cp. Bundesamt für Migration und Flüchtlinge 2005a, Bundesamt für Migration und Flüchtlinge 2007c, Bundesamt für Migration und Flüchtlinge 2008a

49 Cp. Bundesamt für Migration und Flüchtlinge 2009b, 67f.

50 In der Anfangszeit haben wir Themen sicherlich stärker selbst ausgewählt, unter dem Aspekt was wir meinten was relevant ist für die Politikberatung. Es ist in den letzten Jahren aber stärker dazu gekommen dass wir [...] ganz überwiegend mit Aufträgen ausgelastet sind. Die meisten Aufträge kommen aus dem Bereich des Bundesministeriums des Inneren, [...], oder hier direkt aus dem Haus. (Interview with a BAMF researcher, 2015)

51 Boswell 2009b, p. 175

thorough coordination process" in which study proposals can be submitted both internally [...] and externally."[52] "Internally" refers here to other BAMF departments, and not the Research Group itself, which is presented somewhat exclusively as a receiver of study proposals. In the following overview of selected research projects, this impression is further enforced: in the table, every single study is linked to a specific commissioning institution – most often the BAMF and the Federal Ministry of the Interior.[53] It should be noted that this impression was created partly by selectively representing related research projects, partly by relabeling originally self-administered studies as "commissioned by the BAMF," such as the migration potential project.[54] Be that as it may, this evidence indicates that study commissions seem highly valorized by the Research Group. Indeed, in the Research Group, study commissions are regarded as a measure of the degree to which the institution is well-regarded and acknowledged.

Currently, the process of acquiring study commissions is formally organized. Proposals for research projects are collected from other state institutions about once a year through an inter-departmental working group:

> "The inter-departmental working group makes project proposals every year what the BAMF should conduct research on in their view. If an idea is pushed hard, and if it makes sense, or there are two ministries which want something similar, it is taken very seriously. One tries to augment it and turn it into a project proposal."[55]

In practice, research mandates are usually not carried out without further negotiations. All in all, while research proposals are in principal regarded as "good news,"[56] these are scrutinized and become subject to a process of negotiation between the Research Group and other authorities in question.

---

52 Gütlhuber and Schimany 2013, p. 23

53 Ibid., p. 24

54 Field notes, July 2016

55 "Diese interministerielle Arbeitsgruppe darf jedes Jahr Vorschläge machen was das BAMF denn mal in ihren Augen untersuchen sollte [...] Wenn aber für den Vorschlag lobbyiert wird, und der [...] Hand und Fuß hat oder es gibt zwei Häuser, die sich was ähnliches vorstellen, dann nimmt man das schon sehr ernst und versucht dann das anzureichern und einen Projektvorschlag draus zu machen." (Interview with a BAMF researcher, 2016)

56 Field notes, December 2016

"We discuss what has been proposed to us. We check if something exists already on these topics. Sometimes we reject proposals directed to us: if it is not worth the effort, or if it is redundant. Some questions we reject because they are not within our mandate, they are too distant from the BAMF's competencies."[57]

The practice of acquiring study commissions can be embedded into a larger strategy of interpreting the role of the Research Group *like* a departmental research institute despite the lack of legal recognition as such. One expression of this strategy can be seen in the imitation of certain structural features of departmental research institutions. For example, a scientific advisory board was founded in 2005. The advisory board's task was to support the Research Group with expertise and provide its research activities with the legitimacy of renowned academic experts from different migration-related disciplines. The Council's tasks and composition is equivalent to similar organs in departmental research institutions.[58] Another aspect of this strategy concerns the publication of research results: by default, these results are published in a series of research reports, which resembles similar practices of departmental research and can be considered a major novelty in the BAMF, as already mentioned.[59] Furthermore, the Research Group is frequently referring to a catalog of quality standards of departmental research.[60] Also, the Research Group was eager to build up a network of institutional contacts and a working group of departmental research institutions concerned with migration research since 2008.[61] From the perspective of the researchers, this strategy provides a model of reference for its research activities especially towards other state actors, therefore counterbalancing the blurriness of the legal mandate.[62]

---

57 "Wir [...] diskutieren dann was uns vorgeschlagen wurde. Wir schauen dann auch nach, ob es zu diesen Bereichen schon was gibt [...]. Und weisen eben auch Vorschläge, die an uns gerichtet wurden, eben auch zurück: Das lohnt jetzt nicht, das ist Doppelforschung; bei einigen Fragen sagen wir auch das ist nicht unser Auftrag, das ist zu weit weg von dem was das BAMF sinnvollerweise machen könnte."(Interview with a BAMF researcher, September 2015

58 Gütlhuber and Schimany 2013, 27f.

59 Kraler and Perchinig 2017, p. 85

60 Kreienbrink and Worbs 2015, p. 329, Gütlhuber and Schimany 2013, p. 27

61 Bundesamt für Migration und Flüchtlinge 2009b, p. 62

62 This is connected to the fact that, due to constitutional regulations, scientific research cannot be conducted by the Federal Government unless it is directly connected to the executive tasks of government.

In regard to the mission mandate, this strategy achieves two aims: first, it integrates the Research Group into the proceedings of the state bureaucracy; research is not conducted as a result of internal deliberations, but rather as a result of according demand by the BAMF or other state actors. Secondly, this process supports the shaping of an original area of competence vis-a-vis other institutes of knowledge production: this is, for example, visible in the practice of declining research mandates with the argument that they would fit the scientific profile of another institution better. Reference to other institutions, on the other hand, implies a claim to a specific area of migration and integration research which is considered the original competence of the Research Group. The exact confines of this area are not clear from the outset and are subject to negotiation, which will be analyzed in more detail in the next section. The important point in this context is the fact that this area of competence is defined and defended primarily against other departmental research institutions. Quite clearly, the Research Group views itself as the only institution with a clear mandate and competency in migration and integration research:

> "Out of the 47 federal research institutions listed in the federal report on research and innovation in 2012, nine institutions are concerned with migration and integration topics at least implicitly. The explicit primary research focus of none of these institutions, however, lies on these topics. This means that the Research Group at the Federal Office is the only federal institution which is explicitly engaged in migration and integration research."[63]

All in all, the orientation to departmental research is more than just a rhetorical reference, but rather can be considered the core strategy for institutional development for the Research Group. This is well visible in the following passage:

> "In its character, the research of the Federal Office is departmental research, even though it is not formally constituted as a departmental research institute. This means that the research is not primarily theory-driven, but rather application-oriented and provides transfer knowledge. Research does not only provide short-term information for enquiries and statements, but also study projects in the medium term."[64]

---

63 Gütlhuber and Schimany 2013, p. 32

64 Memo from the Research Group sent to the author, February 2014.

According to this reasoning, departmental research is mainly defined in terms of its orientation towards policy-making, and not institutionally, in terms of its position within the administrative hierarchy. Furthermore, departmental research is above all defined by its difference to academic research and is mainly characterized by the fact that it is application-oriented, and not theory-driven.[65] In this view, as a consequence, by fulfilling similar tasks, the Research Group's work is "in its character" departmental research as well.[66] Even though there is a short reference to the lack of legal recognition of the Research Group, this appears rather like a formality and not as an important structural feature.

The most elaborate attempt to define the Research Group's task and to clarify its relationship to the concept of departmental research is contained in the 2015 ten-year anniversary essay:

> "accompanying research in the sense of the legal mandate has to be understood as application-oriented research, which includes the preparation, monitoring and evaluation of policy measures or programs in the area of migration management. This cannot be reduced to the collection of data and information in relation to current or future political measures. For well-grounded 'analytic evidence', this would not be enough. Rather, prognoses are additionally necessary to identify future needs of policy-making. Only the combination of advance and accompanying research in the narrow sense of the word enables us to do justice to the mandate of policy counseling. Departmental research, on the other hand, has an even larger mandate; it elaborates scientific groundwork as a basis for decision-making in administration and policy. [...] Even though the Research Center fulfills these characteristics in part, the institutional set-up is different to departmental research institutions. This creates some differences in matters of finance and personnel [...] as well as the lack of the evaluation through the scientific council."[67]

Again, in this definition of accompanying research, practical relevance and the functional equivalence to departmental research are the two crucial features. In addition, this essay specifies the tasks of research in some detail: practical applicability means that research is an integral part of policy-making and indeed fulfills all functions of scientific policy counseling such as evaluation,

65 Cp. also Bundesregierung 2007, p. 3

66 Cp. also Gütlhuber and Schimany 2013, p. 20

67 Kreienbrink and Worbs 2015, p. 329

monitoring, and prognosis. Again, the influence of the concept of departmental research is obvious since these research goals are almost literally taken over from the Federal Government's research guidelines.[68]

Indeed, the notion of practical applicability seems to serve as a core concept of the Research Group's practice of governmental research. In this context, it is interesting to analyze this notion in some depth: can the provision of practically relevant knowledge serve as a meaningful distinction between the Research Group and other organizations, most importantly academic research institutions?

It is important to note that, as already mentioned, the frequently discussed systematic difference between science and politics is not the most important demarcation line in the institutional arrangement of governmental knowledge production. As discussed above, the Research Group has been quite successful in integrating its knowledge production into the workings of the Federal Office. In contrast to that, it seems that a much more pronounced line is drawn between governmental research and academic or university research. In this logic, not the inherent difference between knowledge production and administration, but the difference between governmental knowledge production and academic research is the most important line of distinction for BAMF researchers. This distinction became apparent, for example, in the following interview passage when discussing theoretic concepts of the BAMF research work:

> "We work flexibly with what serves best. [If] we have a concrete question, we look at which methods we can use to answer the question posed to us. In this we are not overly committed to a specific theoretical concept. If we refer to definitions [e.g. in the National Migration Report], these relate to statistical data, and the statistical data depends on legal regulations."[69]

This does not imply that the Research Group fully disassociates itself with traditions and methods of academic knowledge production, which clearly struc-

---

68 Bundesregierung 2007, p. 3

69 "Wir arbeiten flexibel mit dem was da ist. [...] Wir haben eine konkrete Frage, wir schauen uns an mit welchen Methoden wir die konkrete Frage die uns gestellt wird beantworten können. Und sind nicht übertrieben eng hinter [...] einem Theoriekonzept her. Wenn wir Definitionen benutzen [zB. im Migrationsbericht] richtet sich bei uns nach den statistischen Erhebungen, die statistischen Erhebungen wiederum richten sich nach dem was in unseren Gesetzen drin steht." Interview with a BAMF researcher, September 2015

ture the BAMF's research output: empirical data is gathered and analyzed with scientific methods, the texts use academic literature reference systems, and project results are disseminated in typical academic forms such as working papers, anthologies, or scientific conferences.

Be that as it may, the concept of practical relevance and the partial dissociation from academic knowledge production constitutes nevertheless a basic feature of the Research Group's self-understanding. The reason for this can be found in the strategy of becoming a departmental research institute: according to empirical studies on these institutions, the habitus of dissociation from academic research is discussed as a deeply entrenched feature of departmental research. The general argument goes that the pathologies of academic knowledge production, such as dependency on third-party funds, short-cycle ups and downs in the economy of attention in scientific research, the "publish-or-perish" dilemma and others are referred to as a negative image in contrast to departmental research which is portrayed as "practice oriented."[70] In this context, theory development is not a purpose in itself but rather an instrumental feature of knowledge production: theory is applied to maintain a claim of scientific objectivity, but altered in a way that it fits into "practical relevance" considerations. In governmental research, the preferred mode of operation is the orientation towards a mainstream within a given field to avoid political controversy. Barlösius comments on the structural link between practical applicability and theoretical conservatism:

> "Departmental research delivers good performance if research results stand the test of practical politics. [...] To achieve this, it is rather not necessary, in the contrary even a risk, if departmental research is positioned at the 'peak of science', because these methods and interpretations are often controversially discussed within academy. The use of such research results risks a scientific dispute [...] which could disable political action rather than support it. To minimize this risk it is more favorable to the ministry to use secured, undisputable scientific knowledge and appropriate methods, which are part of the established scientific tool box."[71]

In sum, the orientation towards the model of departmental research as a somewhat neutral provider of practically relevant information to policy-makers lies at the root of this rather over-pronounced disassociation from uni-

70 Barlösius 2008, p. 23

71 Ibid., 15f.

versity research.[72] Indeed, the Research Group puts forward exactly the same concept of governmental knowledge production – on the one hand, commitment to academic method, on the other hand, reference to "practical applicability":

> "We conduct academic studies which apply the usual methods. Only the research question is usually not theory-driven, and that is a difference to universities. Here, we focus on applied research. [There is a wide array of] policy-counseling institutes which likewise follow an academic approach, which are almost always managed by academically trained scientists, but which have a more diversified audience."[73]

The self-understanding as a departmental research institute in form and function can be regarded as the most important factor in the self-understanding of the Research Group.

The requirement of practical applicability systematically influences the knowledge production at the BAMF in various respects: most importantly, this influence is visible in the formulation of the research agenda which is a result of either internal deliberation of "what could be of interest" or the result of a study commission. In regard to methodology and definitions, practical relevance means to use legal definitions and official statistics whenever possible. In regard to theoretical concepts, it stands to reason that despite contrary claims, scientific theory is systematically applied to the texts but usually not explicated. Knowledge production follows the above-mentioned principle of structural conservatism, according to which uncontroversial, mainstream theories and methods are strongly preferred over alternative accounts.

In summary, in the Research Group's definition, the core characteristic of the research work is a specific understanding of practical relevance, understood as a counter-draft to academic knowledge production. In this, differences between accompanying research and departmental research appears rather marginal both in terms of the institutional structure and in terms of the

72 AG Ressortforschungseinrichtungen 2016

73 "Es sind akademische Arbeiten, die genau mit dem ganzen Handwerkszeug arbeiten. Nur ihre Fragestellung ist in der Regel nicht theoriegeleitet, und das unterscheidet sich von dem, was an Universitäten passiert. [...] Bei uns steht die angewandte Forschung im Vordergrund. [Es gibt eine große Bandbreite an] politikberatenden Instituten, [...] die [...] auch akademischen Anspruch haben, die natürlich [...] fast immer von akademisch ausgebildeten Leuten geleitet werden, aber die ein breiteres Publikum haben.Ïnterview with a BAMF reseracher, September 2015

legal mandate. In consequence, the Research Group puts forward a mission statement which claims that it can be considered a departmental research institute in all but the name. In this self-understanding it seems that the institutional make-up is not discussed as a somewhat limiting factor to the research. Quite the contrary, the dependent structure of the Research Group is praised as a specific advantage of the institution:

> "The incorporation of research into the Federal Office and the proximity to the operational tasks of the authority [...] has proven to be a significant factor of success for the work of the Research Group. As a result, synergy effects could be realized and research results with a high degree of practical relevance were produced."[74]

In effect, the mandate of the Research Group in its mission-statement differs little from the original concept of departmental research envisaged by the Independent Commission.

## Institutional Conflict and Cooperation

Up to this point, the structural factors of knowledge production have been analyzed from the point of view of the Research Group: the institutional make-up, the history of foundation, and especially the development of a specific self-understanding around the notion of practically relevant knowledge production have been discussed. In the remainder of this chapter, the relationship between the Research Group and various institutional actors will be analyzed. The relevant institutions include on the one hand what the Research Group regards as peer institutions, namely two socio-demographic departmental research institutions (The institute for Labor Market Research and the Federal Institute for Population Research). On the other hand, they include the vertical institutional hierarchy: the BAMF presidency and the Ministry of the Interior.

Concerning the former, departmental research served as something like an ideal model for the Research Group in the course of its establishment as the most important institutional concept of governmental knowledge production. In practical terms, this model included, besides the acquisition of study commissions, a strategy of coordination with other departmental research institutes, the most important in this regard being the Institute for La-

74 Bundesamt für Migration und Flüchtlinge 2010b, p. 10

bor Market Research (IAB) and the Federal Institute for Population Research (BIB), as mentioned above. The two institutions constitute not only a model of reference, but represent at the same time the most important competitors for resources, study commissions, and epistemic authority. These conflicts illustrate the actual differences between departmental research institutions and the BAMF Research Group quite well. One example in this context is a conflict over personnel between the BAMF and the Institute for Population Research during the foundation phase of the Research Group: the Ministry of the Interior intended to redirect funds to the Research Group at the expense of the BIB, the main argument being the lack of practical applicability of the Institute for Demographic Research's output especially regarding integration policy.[75] A similar situation of conflict arose between the BAMF and the Institute for Labor Market Research in the context of the "refugee crisis" in 2015, when the Institute for Labor Market Research intensified its activities in what was perceived as the original area of competency of the BAMF.[76] In the context of analyzing knowledge production, these incidents are important for two reasons. First, institutional conflicts and struggles within the bureaucracy constitute rather the norm than the exception. However, both conflict situations point to a peculiar asymmetry between the Research Group and its competitors: in both cases, the subordinate role of the BAMF played out as a disadvantage. In the first case, the Institute of Demographic Research was at least in part successful both in warding off the BAMF's attempt to redirect personnel as well as the Ministry's intention to influence its research agenda. In the second case, the Research Group seemed to have too little political leeway to ward off the IAB's "invasion" of the BAMF's territory. Second, it seems that the communication channels between departmental research institutions and the BAMF Research Group seem not developed well enough to absorb such a situation by way of inter-institutional compromise.[77] In any case, this leads to the assumption that the Research Group's institutional network is not very strongly integrated. In fact, regarding the actual research output, there are almost no cooperative publications between the Research Group and other departmental research institutions until 2016.[78] In the Research Group's ex-

75 Bade 2017, p. 68

76 Field notes, July 2016. Cp. also Kleist 2018 who mentions that the IAB has a larger total research output on refugees than the BAMF.

77 Cp. Bade 2017, p. 68

78 Bundesamt für Migration und Flüchtlinge 2016a

ternal relations, the inherent ambiguity of the concept departmental research becomes visible: on the one hand, it serves as the most important conceptual guiding post for institutional development, while on the other, departmental research institutions constitute the most important institutional rivals in regard to research mandates, competencies, and resources. As illustrated above, this rivalry sometimes sparks conflicts of interest which usually play out to the disadvantage of the Research Group. This evidence puts the BAMF's claim to conduct departmental research in all but the name into question. Rather, it seems that the institutional constitution of the Research Group is insufficient to actually play in the same league as its main competitors.

## Strategic Orientation of the Research

If the vertical institutional environment is considered, it becomes clear that the Research Group's efforts to interpret its legal mandate were embedded in a political struggle over the strategic orientation of the research: partly due to the potential use (or threat) of research in the policy process, partly due to the comparatively large area of interpretation left by the blurry legal definition, several forces influenced the role and functions of knowledge production at the BAMF. Key actors in this struggle are, besides the researchers themselves, the Research Group's scientific advisory board, the BAMF leadership, and the Ministry of the Interior.

In 2005, at the time of the establishment of the Research Group, the BAMF was led by a profiled politician, Albert Schmid, a significant difference to many of his predecessors who were usually regarded as administrative specialists with little political ambitions. With his political background, Schmidt recognized research as a strategic tool to enhance the BAMF's area of responsibility vis-a-vis other institutions as well as its political profile.[79] In the years after the institutional restructuring, the BAMF promoted a strategic idea about its new role as a "Competence Center for Migration and Integration"[80]; in bolstering this strategy, knowledge production assumed a key position.[81] Indeed, judging from the visual impression of Research Group publications, the BAMF leadership seemed to attribute increasingly more importance to the Research Group's output: from 2008 onwards, the Research

79 Boswell 2009b, p. 167

80 Bundesamt für Migration und Flüchtlinge 2008d, p. 2

81 Boswell 2009b, 177f.

Group released an annual report about its activities to enhance the visibility and dissemination of its work.[82] From 2010 onwards, the BAMF's president is featured in it with a foreword, stressing the scientific competence of BAMF researchers and the quality of its publications.[83] Resources in terms of personnel and research funds are gradually enlarged as well: from 2007 onwards, research is conducted in three units (formerly two); from 2010 onwards, the Research Group receives a flat-rate research budget (in addition to project-based funds and staff).[84] The name of the research unit changed from the initial designation "Research Group" to "Research Center for 'Migration, Integration and Asylum'" in 2014, reflecting the ambition to provide it with a more publicly visible profile.[85]

However, the growing importance of research from the BAMF leadership's perspective had some bearings on the research agenda, too. By and large, the BAMF favored an application-oriented, demand-based orientation of research, a concept which had to be implemented at the expense of more general, long-term foundational research. In the leadership's concept, research was less a source of knowledge relevant for decision-making but rather a means of symbolic authority, to demonstrate expertise in the field of migration and integration to the public and other authorities. While this strategy effectuated a higher visibility of the Research Group and its work, it came at the cost of rather menial non-research tasks such as speech-writing, briefing of higher BAMF officials or memo compilation.[86] In the BAMF leadership's concept, the development of an academically oriented scientific profile of the Research Group was clearly of secondary importance.

In contrast to this, the scientific advisory board of the Research Group favored a more independent role of research with a focus on long-term research projects. The advisory board's idea about research tasks can thus be seen in line with the Independent Commission's concept of the role of research. These diverging ideas about the long-term strategic orientation of research were the

82 Bundesamt für Migration und Flüchtlinge 2009b

83 Bundesamt für Migration und Flüchtlinge 2011c, p. 9

84 Bundesamt für Migration und Flüchtlinge 2012f, Gütlhuber and Schimany 2013

85 Kreienbrink and Worbs 2015, p. 326. To avoid confusion, this text uses the term "Research Group" in a consistent manner; this follows the nomenclature of most BAMF-researchers who refer to this expression in interviews.

86 Kraler and Perchinig 2017, p. 75

subject of frequent discussions between the BAMF leadership and the scientific council, as one member of the council describes:

> "In [the BAMF president's] view this institution should above all conduct commissioned research. According to his definition this meant 'giving answers to posed questions'. The researchers were constantly overburdened with tasks like briefings and speech drafts. So we negotiated successfully to grant more freedom for independent scientific work."[87]

The Ministry of the Interior as a superior authority played a rather unclear role during the Research Group's early years: in the course of the drafting of the Foreigner's Act, the ministry was one of the driving forces to limit the impact of the Independent Commission's proposal of knowledge-based policy-making, which ultimately effectuated the low degree of the Research Group's institutional independence. Hierarchically, the Ministry of the Interior is responsible for the *Fachaufsicht*, the administrative, technical, and legal supervision of the Research Group which entails (at least formally) influence on the research work. The ministry's control reservation is frequently cited (usually off tape) as an important impediment to a more independent general research strategy.

> "It would be possible to change the residence law and delete the research paragraph, or it could be interpreted differently: 'research means that one employee compiles some information for the federal government.' It is a very flexible term. That means there is a certain dependency on the good-will of the Ministry of the Interior."[88]

---

87 "Der BAMF-Präsident vertrat die Auffassung, dass diese Institution vor allem Auftragsforschung übernehmen solle. Nach seiner Definition hieß das: ‚Antwort geben auf gestellte Fragen'. Die Forscher wurden dadurch und durch viele andere Aufgaben, z. B. durch Briefings und Entwürfe von Reden, zeitlich überfordert und zuweilen auch qualitativ unterfordert. Deshalb haben wir – erfolgreich – darüber verhandelt, ihnen mehr Freiraum für die selbständige Forschung einzuräumen." (Interview with a former member of the BAMF advisory board, 2017)

88 "es wäre ja möglich zu sagen, das Aufenthaltsgesetz wird geändert und der Forschungsauftrag wird gestrichen. Oder der Forschungsauftrag wird ganz anders ausgelegt. Der Forschungsauftrag besteht darin, dass ein Mitarbeiter für die Bundesregierung ein paar Informationen zusammenstellt. Das is ja ein sehr dehnbarer Begriff. Das heisst es gibt natürlich eine Abhängigkeit von der Gunst der Leute, die im BMI sitzen."((Interview with a BAMF researcher, 2016)

However, after establishment, the Ministry did not directly interfere with the research work of the BAMF, except for the exercise of the usual routine of editing report manuscripts.[89] Some interview partners as well as research literature mention that in the initial phase, the Ministry of the Interior had in fact little use for research, which would at a first glance contradict the assumption of the exercise of control.[90]

However, when commenting on the role of the ministry in interviews, a common theme consists of the rather discrete and indirect mode of control, as is, for example, visible in the quote above; the ministry's influence is not easy to pin down directly in certain restrictions, actions, or hierarchical orders, as in the case of the BAMF leadership. Rather, its control reservation is experienced as a "feeling of dependency on the good will." Evidence from academic research indicates that these indirect control tactics can be found in the Ministry's conduct vis-a-vis affiliated political actors, such as the Independent Commission or the Islam Conference. Research on these political bodies reveals relatively subtle forms of control via "paper technologies"[91] which remain shy of the level of what is considered undue political influence. These forms of influence include, in the case of the Islam Conference, the BMI's production of conference session protocols. These protocols were presented as a service to the conference but were produced slightly biased towards the positions of the state in terms of length, accuracy of representation, and plausibility.[92] In the case of the Independent Commission, the ministry exercised tacit influence on the production of internal papers by way of affiliated staff in the Independent Commission's office as well. [93] In the case of the Research Group, indirect methods seem to prevail as well. This can be illustrated with the following interview passage describing the process of creating an interdepartmental working group responsible for drawing up proposals for future BAMF research projects. This was preceded by an unusual accumulation of complaints about the BAMF's research topics and publications:

> "The problem was that the supervision control at the BMI let the Research Group's notices heap up on his desk and did not forward them to other ministries which worked on similar research questions. This resulted in unnec-

89 Kreienbrink and Worbs 2015, p. 330

90 Boswell 2009b, p. 175

91 Engler 2018 forthcoming, p. 64

92 Ibid., 260ff.

93 Schneider 2010, p. 265

essary redundant work, which led to a request for additional administrative control. This in turn complicated the Research Group's work even further."[94]

In the quote, it seems that the ministry's alleged lack of attention effected tightened control via a newly established supervision body over the Research Group. In the case of the inter-departmental working group, this subtle control is not the consequence of an all-encompassing, "Foucauldian-style"[95] surveillance strategy with indirect means; rather, it can be regarded as a more or less random result of neglect of bureaucratic duties.

The point is, however, that this neglect played out in favor of the Ministry by enlarging its grip on the Research Group's inner workings.

The most important entry point for indirect influence can be discerned in the practice of commissioning research as seen in the case of a study on naturalization. As a result of the reform of citizenship law in 2000, children of foreign nationals acquired German citizenship by birth but were in some cases required to discard the foreign nationality before reaching the age of 23, otherwise the German citizenship would be revoked. In theory, the so called *Optionspflicht* ("mandatory option") regulation was meant to reconcile the *ius soli* concept of citizenship with the foundational principle of restricting access to dual citizenship wherever possible. However, in practice, dual citizenship was more a rule than an exception since in about half the naturalization processes the other citizenship was not revoked.[96] As a result of exemptions of EU nationals, the mandatory option regulation targeted mostly Turkish nationals, which sparked considerable political criticism with charges of discrimination and an undue bureaucratic harassment of prospective future citizens. In reaction, by commission of the Ministry of the Interior, the BAMF conducted two studies on the effects of this unique and controversial legal regulation.[97] The results of the studies were rather positive: not only did

94 "Das Problem lag darin, dass hochrangige Beamte des BMI, denen die Fachaufsicht oblag, Meldungen der Forschergruppe nicht weitergaben an andere Ministerien, in denen zum Teil ähnliche Forschungsfragen delegiert oder diskutiert wurden. Das führte zur Vorstellung von unnötiger Doppelarbeit und zu der Forderung nach administrativer Kontrolle der Forschergruppe des BAMF, was deren Arbeit unnötig erschwerte." (Interview with a former member of the BAMF advisory board, 2017)

95 Boswell 2011

96 This applied to all EU citizens, as well as citizens of countries which were unusually uncooperative in releasing their subjects from citizenship.

97 Bundesamt für Migration und Flüchtlinge 2012d and Bundesamt für Migration und Flüchtlinge 2012b

almost every young foreigner opt for German citizenship, but a large majority reported that the decision was rather easy for them. Consequently, the study results were regarded as proof of the success of the policy by the Ministry for the Interior.[98] However, these studies were conducted on a group of people born between 1990 and 2000 whose parents applied for the mandatory option as part of a transitional arrangement. This decision was justified with logistic necessities, since this special group were the only people who could be included in such a study since the law was otherwise only applied to newborns which were at that time too young to be included in a social survey. However, from a methodological point of view, this selection created a source for bias. The administrative hurdles to accessing the mandatory option were somewhat higher for this group, since parents had to file an application and pay a fee of 500 Marks. Indeed, another study presented data which points to a positive selection in terms of socio-economic data, German skills, and support for German citizenship by the social environment of this sub-group; given the legal requirements of fee and formal application, these findings are not overly surprising,[99] as a member of the Research Group's scientific advisory board confirmed:

> "The parents take an interest in the German citizenship. They will tell this to their children, and will comfort them if they have identity conflicts. The result of this, as we argued, will be the information to the public that everything was allegedly completely unproblematic. And this is exactly what happened: the Federal Secretary of the Interior proudly presented the successful and unproblematic implementation of the legal regulation."[100]

This study illustrates nicely how different political actors exercise political influence on the generation of knowledge: study results can be framed and influenced in a particular direction, as in the case of the dual citizenship

98 Bax 2012

99 Diehl and Fick 2012, p. 349

100 "Die Eltern haben ein Interesse an der deutschen Staatsangehörigkeit ihrer Kinder. Sie werden ihnen das nachdrücklich deutlich machen. Und wenn die Kinder beim Wechsel der Staatsangehörigkeit irgendwelche Identitätsprobleme haben sollten, werden sie sie ermutigen. Das Ergebnis wird die öffentliche Information sein, so haben wir damals argumentiert, dass das alles angeblich völlig unproblematisch sei. Und genau das ist dann passiert: Bundesinnenminister Friedrich berichtete stolz von einer erfolgreichen und ganz unproblematischen Umsetzung der Regelung." (Interview with a former member of the BAMF advisory board, 2017)

study through selection of a non-representative sub-group. The researchers are aware of this problem and react with detailed, methodologically sound discussions of the potential bias sources of the sample data and frame the study in a sober, unassuming way. This methodological discussion, however, is not part of the political communication strategy. The detailed documentation of bias sources is relegated to the background of a "success story" based on the finding that most migrant youth opt for the German passport without much inner conflict. With this, the BAMF's study is interpreted as a support of the politically contested regulation without mentioning the fact that the rather positive study results can be attributed to a large part to the bias created by the study's target group. The policy in turn is supported by the allegedly objective, sober scientific findings.

This story points to another mechanism in the way knowledge is produced on politically sensitive topics in general. In the case of the study on naturalization, the Research Group seemed to be keenly aware of the political controversy and the possibility of exploitation of study results for different political purposes. In line with the Office's general defensive policy in terms of public relations, the researchers adopted a role of impartial providers of information by refraining from overt recommendations in this study:

> "some of our studies draw conclusions on what could be done. But in such a contested area like for example citizenship [...] we didn't do that. We made a proper study, we analyzed [different] effects and presented [the material]. If you look at the [...] press releases [of two contesting political actors, V.K.], one could think they referred to two different studies. But [...] we were OK with that, because everyone can work with this material."[101]

By adhering to the "neutral presentation of scientific facts" on especially sensitive political topics, the Research Group interprets its role as a source of information for all political parties.[102] From the point of view of the Research Group, this communication strategy is sensible since it avoids criticism of partisan knowledge production which could undermine the credibility of scientific results and, eventually, the carefully constructed reputation of the Research Group as a quasi-departmental research institution. At the same time, this kind of knowledge answers to the given demand, as Amir-Moazami argues:

101 Interview with a BAMF researcher, September 2015

102 Kreienbrink and Worbs 2015, p. 330

> "A correlation materializes which seems both obvious and paradox at first glance: the demand of rationality rises in the same rate as does the politicization of the field which renders objectivity and neutrality basically impossible."[103]

In the case of the Research Group, this communication tactic points to a certain ambiguity of the strategy of acquiring research mandates: on the one hand, study commissions ensure institutional recognition and are seen as proof of practical applicability. On the other hand, it provides an entry point for political actors to influence the research agenda without compromising the scientific credibility of research results. At the same time, the strategy of producing objectivity by focusing on methodologically sound research combined with a decidedly defensive communication geared to avoiding criticism could arguably further enhance the political actor's possibilities to exploit study results in a partisan way. By defining "scientific neutrality" in a way that study results can be used by all political actors to bolster their respective and usually contradicting political claims, the study results become in a way random and prone to arbitrary interpretation.

In conclusion, empirical evidence of different strategies of control and influence over the research agenda were described as a framework of institutional preconditions for the formulation of knowledge. This framework is on the one hand somewhat typical for institutions of governmental knowledge production, as the frequent reference to departmental research institutions shows. On the other hand, most importantly the blurriness of the Research Group's legal mandate represents a somewhat unique situation which has been explored in some detail. The process of institutionalization can be characterized as the result of a struggle over the research strategy between the Ministry of the Interior, the BAMF leadership and the Research Group itself. The actors represented different ideas of how governmental research ought to be oriented. The Research Group and its Scientific Advisory Board were eager to shape its institutional make-up like that of departmental research institutions, an approach which entailed most importantly a strategy to acquire commissions for the systematic inclusion of research results in the process of policy-making. The BAMF leadership, on the other hand, favored a different role of the Research Group as having a much lower scientific profile – compiling memos, speeches, or short studies for ministerial requests

---

103 Amir-Moazami 2018b, p. 111

instead of conducting foundational research. In the case of the ministry, the main characteristic of its control practice is the absence of direct influence on the research process itself – the selection of methods and analysis modes seems to be a truly independent area of decision for the researchers.[104] By focusing on tacit, indirect methods of influence, both of the ministry's roles as described in interviews – disinterestedness and control reservation – are less of a contradiction than expected at first glance.

## Conclusion

In this chapter, the reforms of the "paradigm change" has been recapitulated from the point of view of the Research Group as an involved actor. In this context, the notion of a "paradigm change" was crucially connected to a new role of knowledge in the process of policy-making in migration politics as suggested by the Independent Commission Immigration: scientific knowledge promised more reasonable, effective and coherent politics. In this respect, it seems plausible to assume that the Research Group represents this paradigm change like no other institution in the current set-up of German migration policy-making. However, many of the Independent Commission's proposals especially in regard to knowledge production were sacrificed in the legislative negotiations. In this context, the blurriness of the legal mandate was not a somewhat accidental result of the parliamentary process surrounding the implementation of the Independent Commission's recommendations. Rather, it can be regarded as a result of a strategy to systematically diminish the institutional influence of research in policy-making as a whole: this strategy is materialized in the removal of the Immigration Council, the inclusion of knowledge production into the BAMF hierarchy, and the resulting political primacy over the research agenda. In reference to the original proposals put forward by the Commission, the selective implementation of proposals display a bias towards administrative control, thereby strengthening especially the central role of the Ministry of the Interior. This strategy did not only refer to the reform elements discussed here, but can be considered the general pattern of implementing the Independent Commission's reform proposals.[105]

104 Barlösius 2008, 17 f.
105 Schneider 2010, p. 635

Taken together, the Research Group's establishment can be regarded as a complex interaction between political actors and their different visions about the role of knowledge production in politics on the one hand and strategies on the other to give meaning to the rather blurry legal mandate of accompanying research at the BAMF. In this situation, the Research Group developed a strategy of imitating a departmental research institution, therefore providing a model of orientation for its research activities. As a result, the Research Group developed a self-understanding that it conducts departmental research in all but the name; in a way, this claim can be read as the fulfillment of the Independent Commission's vision of a knowledge-based migration and integration policy. A measurement of success of this strategy can be found in the fact that initial tensions and alienation between the research staff and the administrative units of the BAMF seem to have largely given way to a smooth integration of the different units of the authority. However, this success is somewhat put into perspective by the fact that both the BAMF leadership and the Ministry of Interior took advantage of the low degree of institutional independence of the Research Group and exerted considerable influence on the strategic orientation of research. In interviews, the political aspect of control is usually mentioned in explaining this specific formal set-up; there seems to be a consensus that the interest of political control outweighs the merits of a truly independent research institution from the perspective of the ministry. As a result, the Research Group is subject to two different supervision hierarchies and rationales: the ministry was limiting the potential political problems arising from independent research by confining the institutional independence of the Research Group with indirect means, while the BAMF was eager to turn the Research Group into a productive factor for its political strategy. Not incidentally, both institutions are the most important study commissioners to the Research Group. Through this practice, both the BAMF leadership and the ministry have the means to crucially influence the knowledge produced by the Research Group without directly intervening with the research process and therefore undermining the scientific credibility of the study results.

In the institutional make-up, some evidence can be found which illustrates how this political influence materializes in terms of the organization of research. This is expressed by the fact that, staff and budget wise, the Research Group still stagnates and has since about 2013. Initially, personnel as well as financial funds were constantly expanded, but these reached a peak around 2013 of about 25 scientific staff and ca. 400,000 EUR, respectively.[106] Another aspect of the situation of the Research Group is the fact that most of the job positions are limited to two-year periods which limits the attractiveness of the work conditions as well as the ability of individual scientists to develop expertise in their given field of responsibility.[107]

All of these factors combined point to the structural limits of research in its present form:

> "The researchers successfully put a lot of effort into producing sensible results within the confines of the possibilities presented to them. However, research would be much better if the group was larger, if it was managed according to scientific principles, and if its research would be better integrated strategically."[108]

As a result, while the Research Group maintains that it conducts departmental research for all practical concerns, it stands to reason that the structural confines are more limiting than conceded in the official mission statement. One effect of this institutional dependence is the Research Group's defensive approach to political controversies and the resulting restraint in political recommendations. Instead of providing knowledge to inform and monitor political measures, research results are distinctly formulated in a way to avoid evaluation of political measures, motivated mostly by the fear of avoiding public criticism of partiality. This effect is not, however, random or a standard feature of all Research Group publications. Rather, the restraint in terms of policy recommendations is greatest in politicized issues. Ironically, these politically heated questions were a prime target of scientific knowledge production in the concept of the Independent Commission to begin with: ideologically framed policy fields subject to a decade-long stalemate and reform backlog were to be reformed by superior technical knowledge. In precisely these policy areas, however, research results are formulated in a way that they can be applied to support almost any political claim. Knowledge production is no longer

---

106 Email from the Research Group, February 2014

107 Field notes, december 2016

108 "Die Forscher bemühen sich im Rahmen ihrer Möglichkeiten erfolgreich darum, vernünftige und tragfähige Ergebnisse zu produzieren. Diese Möglichkeiten wären aber deutlich besser, wenn die Anlage größer wäre, wenn sie wissenschaftlich klarer geführt und forschungsstrategisch besser eingebettet würde." (Interview with a former member oft eh BAMF advisory board, 2017)

a basis for policy-making in the sense of an external, to a degree independent, voice which is in the position to monitor, evaluate, or give recommendations to policy. In a way, the relationship between policy-making and knowledge production is almost the reverse of the Independent Commission's concept.

# Analysis of Governmental Knowledge Production

In the last chapters, the structural conditions of governmental knowledge production have been outlined: the Research Group has built up a narrative of governmental knowledge production on migration which is shaped after the image of departmental research and a concept of instrumental knowledge as put forward by the Independent Commission. In the last chapter, the structural conditions and practical constraints of governmental research at the BAMF were analyzed: contrary to the usual practice, the research unit was not set up as an independent departmental research institution but incorporated into the administrative structure of the Federal Office. In the first years of its existence, the Research Group attempted to carve out a secured area of competence within these confines, mainly by mimicking structural features of departmental research institutes. While this strategy was successful to a degree, it also constitutes a main entry gate for political manipulation of research results. Furthermore, the mission statement of the Research Group – providing knowledge for informing policy-making – is in practice severely restricted by the quite peculiar institutional arrangement of knowledge production. The main outcome of the analysis of the structural conditions is that the research work is characterized by an unresolved conflict between an ideal role of knowledge-based policy-making on the one hand and the institutional restraints on the other. All in all, the degree of scientific independence correlates negatively with the potential of political conflict attributed to a specific research topic.

Bearing these structural conditions as well as the self-perception of the Research Group in mind, in this chapter the published research output produced by the BAMF will be analyzed. The main focus lies in explaining how the institutional and intellectual framework dynamically interacts with the knowledge produced in the BAMF: basically, this interaction shapes a specific governmental knowledge subject to dynamic changes over time. The analytical tool applied here is a typology of four ideal types of knowledge-power complexes[1] which include the main topics of research and their intended political use. This chapter is divided in two parts: in the first step, the typology of BAMF publications is developed. By using methods of lexicometric analysis, an overview of methods, topics, and theoretical concepts of all BAMF research publications is created. This overview is completed with an analysis of the notion of practical relevance as a core defining feature for knowledge production at the Federal Office. In the second part of the chapter, the four ideal types of knowledge-power complexes are analyzed which connect typical topics of research with associated practices of knowledge production as well as practical knowledge considerations: These include first, knowledge for administrative purposes using the example of Migration Reports; second, depoliticizing knowledge with the example of integration studies; third, defensive knowledge given the example of studies on Muslims; and fourth, legitimatory knowledge with the example of knowledge about African and Eastern European migration.

1 Cp. Amir-Moazami 2018 p. 92ff.

## Framework of Analysis

As a first step towards an analysis of the knowledge produced at the BAMF, it is worthwhile to approach the material with a variety of quantitative analysis methods. Due to the size of the corpus of literature altogether containing thousands of pages of text, such a step is necessary for a preliminary overview. To achieve this, the Research Group's own representation of its work is presented first with a focus on the internal research organization and the range and selection of research topics and methods. The main sources of this self-portrayal are yearly reports and public relations material.[2] In the second step, this information is analyzed against academic research to highlight differences and similarities between the BAMF's and academic research output. The chapter concludes with a preliminary typology of BAMF research projects.

## Quantitative Overview

As a first step, the base sample of research publications has to be defined. How can research be separated analytically in a meaningful way from the multitude of publications issued by the BAMF? In fact, in the course of its existence, the Research Group has produced hundreds of documents in different forms. Some publications were taken over from other government authorities; others were compiled by contracted authors who were not BAMF officials. To make a useful distinction between what counts as a document of governmental knowledge production and what does not, the BAMF's own concept can be used as a point of departure. For the Research Group's ten-year anniversary, the BAMF compiled a publication list which will be used as a basis for document analysis. According to this list, the Research Group has published the following texts in the ten years between 2005 and 2015:

- 65 Working Papers
- 26 Research Reports
- 8 texts from the *Beitragsreihe* ("Publication series")
- Yearly Migration Reports
- Yearly reports of the Research Group (since 2008/2009)
- Yearly European Migration Network (EMN) policy reports

---

2 Cp. Bundesamt für Migration und Flüchtlinge (Ed.) (2008d), Bundesamt für Migration und Flüchtlinge (2010b)

On the BAMF website, these publications are grouped into three categories:

- **Migration Reports** provide a comprehensive overview over the yearly development of in and out-migration;
- **Research reports** provide an outlet for the publication of larger research projects of the Federal Office for Migration and Refugees;
- **Working papers** contain contributions of the Research Group's scientific staff concerning either preliminary results of larger research projects or self-contained smaller studies.

These three in-house publication series constitute the total body of literature for the analysis here. This selection excludes two sets of publications: first, expert opinions, and second, volumes of the "Publication Series."[3] Concerning the first type, expert opinions differ from all other publications since these texts are not produced by Research Group members and were therefore excluded from quantitative analysis. As for "Publication Series" texts, the authorship question is different since at least parts of these publications are written by BAMF researchers. In general, the texts are visibly directed towards an academic audience: some texts in this series are academic qualification theses, while others are anthologies based on academic conferences organized by the BAMF and presented in the style and form common for academic publications. However, these texts are marked as "not representing the institutional opinion of the BAMF" and can thus be considered outside of the "official" body of literature. This exclusion can be justified with what Iver Naumann called "a bureaucratic mode of knowledge production": in his ethnography of the Norwegian Foreign Service, he observed a high degree of conformity and redundancy in the texts and speeches produced by diplomats; a common characteristic was the almost complete elimination of personal notions.[4] Following this argument, the note that a text "does not represent the institution's opinion" is a very strong argument to not regard it as part of official BAMF knowledge since it stands to reason that the "Publication Series" and expert opinion texts are subject to different production rationales than those outlined in chapter 3.

3 In some cases, both publication types were nevertheless used as sources for the qualitative analysis in the second part of this chapter.

4 Neumann 2012, p. 86

As a first approach to the research texts, the total corpus of the BAMF research output[5] was quantitatively analyzed using corpus-linguistic analysis.[6] In the first step, an n-gram analysis was conducted. N-grams are combinations of words (2-gram = 2 words, 3-gram= 3 words, etc.) which are analyzed according to their relative frequency in the text. N-grams are a relatively basic and, to a degree, simplistic method of corpus-linguistic analysis: for example, while the frequency of "Persons with a Migrant Background" can be compared with the frequency of "Persons without a Migrant Background," it is impossible to determine contexts of meaning, especially if contexts are not manifested in words that appear right next to each other in the text.[7] The n-gram analysis was conducted with the N-gram statistical package, an open-source software using PERL code. The resulting n-grams were combined with bibliographical data of the respective publication in an access-data base. After sorting and refining the data, a list of ca. 7,000 1-Grams which constitute the most frequently used words in the BAMF research was created. The top-10 entries are:

a) Year (*Jahr*)
b) Germany (*Deutschland*)
c) German (*Deutsch*)
d) Foreigner, foreign, foreign country (*Ausländer, Ausländerin, ausländisch, Ausland*)
e) Person (*Person*)
f) Illegal, illegality (*illegal, Illegalität*)
g) Respondent, ask, question (*Befragen, Befragt, Befragte, Befragung*)
h) Migrant (*Migrant, Migrantin*)
i) Other (*andere*)
j) Woman, female, share of females (*Frau, weiblich, Frauenanteil*)

The keywords reference broadly to three areas of knowledge: first, empirical social research (visible in terms such as respondent, question, person); second, the nation state (Germany, foreigner, illegal); and third, related to

5 According to a 2015 publication list, see Bundesamt für Migration und Flüchtlinge 2015c
6 Bubenhofer, no date
7 Rosenberg 2013

both previous items, categorizations of people (such as migrant, woman, and other).

This list was analyzed for frequently used topics as well as topical shifts over the year. However, the limits of the analysis became apparent: in general, as already mentioned, quantitative n-gram analysis is restricted to words or phrases instead of content; furthermore, the discursive meaning and impact of a specific term does not automatically correlate with the frequency of its use. The distribution over time could give insight into a development of ups and downs in certain research topics, but in this respect the BAMF literature proved to be relatively stable: most of the terms mentioned above retain their relative frequency over the years. All in all, n-gram analysis rendered some general trends in the body of literature visible, most importantly the joint administration-scientific speaker perspective which will be analyzed later in some detail.

For a thorough quantitative overview of the Research Group's topics and methods, the yearly reports provide a useful point of departure. These contain some information on how the BAMF presents the organization of its knowledge production. In relation to its research topics and methods, the Research Group is keen on drawing an image of a high degree of versatility and diversity of the research work.

> "The research center is characterized by the fact that very diverse methods are used here, since very different projects are being conducted. There are projects in quantitative sociology which last for four years and at the same time descriptive desk studies, conducted by politologists which are ready within three months. In sum, they span over the whole range of qualitative and quantitative research."[8]

This diversity is related both to the nature of the research interest and the inclusion of the Research Group in the institutional hierarchy. In part, this narration draws on a common conceptualization of academic migration re-

8 "Das Forschungszentrum zeichnet sich dadurch aus, dass hier die diversesten Methoden benutzt werden, das es die verschiedensten Projekte bearbeitet, also von Projekten [im Bereich quantitativer Soziologie] die vier Jahre dauern, und genauso deskriptive Desk Studies, von Politologen bearbeitet, die innerhalb von drei Monaten fertig sind. Also sie decken die gesamte Spannbreite an qualitativer und quantitativer Forschung ab. (Interview with a BAMF reseracher, September 2015)

search as being a thoroughly "interdisciplinary" object of scientific enquiry.[9] Additionally, the Research Group fosters this claim by the collaborative, demand-driven method of compiling the research agenda. In this context, the argument goes that, since most studies are commissioned by other government agencies, there is little leeway in narrowing down the range of research topics:

> "In contrast to theory-driven academic research, the research questions are based on the interests of the contractor. In this regard, the Research Center has not formed topical focus areas, since usually, upon completion of a study, a new project with a wholly different content waits."[10]

Research work at the BAMF is organized in short to mid-term research projects. As of 2015, the Research Group completed 92 research projects (including ongoing projects) while 33 projects have apparently never been concluded.[11] With the exception of aborted projects, research projects are usually connected to a specific publication, either as part of the Working Paper or Research Report publication series, respectively.

Research projects are sorted into topical categories. In general, the Research Group differentiates between the categorizations of migration and integration research, respectively: while integration research represents a unified category, migration research is further divided into several sub-categories:

Migration research

- General aspects of migration (17 projects, including 10 Migration Reports)
- Worldwide migration movements (30 projects)
- Migration and labor market (23 projects)
- Demographic aspects of migration (8 projects)
- Irregular migration (9 projects)

Integration research (38 projects)

---

9 Cp. for example Mecheril et al. 2013, 13f.

10 Kreienbrink and Worbs 2015, p. 329

11 Data from Research Group yearly reports, own survey. All following data refers to the base of 126 research projects and 109 publications, respectively.

Over the years, for the most part, this topical structure was relatively stable, occasional renaming of several sub-categories notwithstanding.[12] One exception to this rule is the topic of demography, which was listed as a separate category along with migration and integration research in the 2007/2008 report. In all successive reports, the category became part of the general migration section.[13] This restructuring is an expression of a general trend of diminishing research activities in this area: since 2009, no new projects have commenced in this area; furthermore, 3 out of 8 research projects have been erased from the research agenda in more recent yearly reports. Since no corresponding study has been published, this indicates most probably that these projects have never been concluded.[14] A similar trend can be observed in research of irregular migration: out of the nine projects in this area, five were concluded before 2008 and the remainder is made up of studies conducted within the framework of the European Migration Network (EMN). Diminishing research activities in these two subjects is counterbalanced with increasing efforts in worldwide migration as well as labor market research. Both topics feature a rising trend among the newly commenced projects from 2011 on. Integration research as the single largest research field does not follow a distinct trend and maintains a stable and relatively high share of the newly commenced projects throughout the years.

In regard to research topics, it is interesting to compare academic migration research with the research output of the BAMF. If the list of research projects and topics is compared with peer institutions, some differences and similarities between academic and governmental research become visible. According the research literature data base SOLIS and the research project data base SOFIS, migration research in Germany is a mid-sized topical area contributing about 5 percent of all German language social science research projects between 1998 and 2008.[15] Within the field, there is a high concentration of research projects at a small number of institutes: around

---

12 For example, "illegal migration" was renamed into "irregular migration" in 2009.

13 "*Themenschwerpunkt Demografie*" Bundesamt für Migration und Flüchtlinge 2009b, p. 68

14 Bundesamt für Migration und Flüchtlinge 2016b, own survey.

15 All numbers in this paragraph refer to the years between 1998 and 2008 and are quoted after Schimany and Schock 2012. See Leibnitz-Institut für Sozialwissenschaften 2010, 12f. for a detailed description of the databases and data collection methodology.

two-thirds of all research projects have been conducted by the top one-third of the listed institutions; among them, only six have conducted more than 30 projects.[16] Compared to this group, the BAMF can be considered a rather large research organization in terms of publication volume and frequency. In 2010, the BAMF was the fifth-largest publisher of migration research in the SOLIS database. However, it should be noted that the numbers refer to the period between 1999-2008, while the Research Group only started publishing in 2005.[17]

If the research topics of academic and BAMF research are compared with each other, some similarities and differences become visible. Concerning academic publications according to SOLIS data, the database lists the following five topics as most relevant:

a) Racism/Discrimination (*Rassismus/Diskriminierung*), 12.1%
b) Multiculturalism/Ethnicity (*Multikulturalismus/Ethnizität*), 8.3%
c) History of Migration (*Geschichte der Migration*), 8.1%
d) International Migration (*Internationale Migration*), 6.6%
e) Circumstances of Life/Education (*Lebenslagen und Bildung*) 6.5% each

If the BAMF research agenda is compared against this list, some differences and similarities are discernible: a distinct focus on socio-economic integration – expressed in topics like education, circumstances of life, and socialization – is the main similarity between academic and BAMF research. Within this topical area, however, the BAMF has rather different focus points: regarding education, the BAMF research is less focused on the general school system, as it is the case with academic migration research. Instead, BAMF research is rather concerned with integration-policy related education, most prominently in the context of the so-called integration courses.[18] Another re-

16 According to Schimany and Schick 2009, 22f., these institutions include the Center for Turkish Studies (Essen), the Institute for Interdisciplinary Research on Conflict and Violence (Bielefeld); the Institute for Migration Research and Intercultural Studies (Osnabrück); the Berlin Institute for Comparative Social Research; the European Forum for Migration Studies (Bamberg); and the Swiss Forum for Migration and Population Studies (Neuchatel/Switzerland).

17 Cp. Leibnitz-Institut für Sozialwissenschaften 2010, p. 26

18 Cp. Bundesamt für Migration und Flüchtlinge 2014b, p. 135: the analysis of the integration of immigrating spouses is partly designed to evaluate federal integration policies, above all the integration courses.

markable difference is the BAMF's approach to labor market research, which is subsumed under the migration section of research, instead of integration as it is usually the case in academic research.[19] This deviation from the rest of the research field can explained with the practice of acquiring study commissions: many study commissions in the field of integration originate directly from the BAMF, which explains why research is less oriented towards the school system and more towards federal integration policies.

There are some differences between the BAMF research agenda and the academic field as well. Some topics which feature prominently on the BAMF's research agenda seem to be less relevant in the academic world: this is above all true for demographic research, which is a comparatively small research topic in the SOLIS data (2.7% of all migration research publications in this area), and to a degree research on irregular migration, which does not appear as a topical category at all, indicating its comparatively low significance.[20] Both topics have, however, constituted a major research focus at least in the first years after the establishment of the Research Group. On the other hand, the almost complete lack of BAMF research on racism/discrimination and multiculturalism is a striking difference to the academic field: the two most important research topics among academic publications have not been the focus of *any* BAMF research project; indeed, they have only been occasionally mentioned in BAMF publications at all.[21]

In terms of project length, research projects are labeled either "new" or "ongoing" for an average of about two years; exceptions include projects which represent standard yearly publications, namely the "migration report" and the "integration report." A related issue are multiple publications which deal with the same project topic such as "Migration Potentials," "Integration of Asylum Seekers," and "Integration Panel"; these projects lasted for two, three, and six years, respectively. Another type of long-lasting research projects consists of empiric studies for which data has to be collected before the actual analysis; most of these studies deal with integration topics. With an average of almost three years, projects in the topic area of demography last for a longer-than-average duration as well. On the other end of the scale are all studies in con-

19 This difference will be discussed in some depth later in the text in the analysis of "Migration potential" studies.

20 Numbers quoted after Leibnitz-Institut für Sozialwissenschaften 2010, p. 20

21 For example, the yearly reports mention three (of more than 500) external publications in this topical area between 2008 and 2015. (surveyed by the author)

nection with the European Migration Network (EMN), called "short studies" or "focus studies," respectively: these entail about one quarter of all research projects (33 out of 126) and almost all concluded within one year. Many research projects in the area of "worldwide migration" are comparatively short as well.

In conclusion, the self-portrayal of the BAMF's research output is consistent with the understanding of research as a provider of practically relevant information to the administration and the political system. This understanding is visible, for example, in the claim that the BAMF's research is not reduced to single topics and can be considered methodologically versatile. This is linked to the demand-driven mode of research agenda-setting. In terms of research topics, there is a distinct shift visible over the years away from knowledge production on demography and illegality towards labor migration and international/worldwide migration, while integration research maintains a high level of research activities over the years.

There seems to be a relatively clear distinction of at least two types of studies in regard to methodology and topic; certain standard types of research designs are used over and over again. A first type is what the BAMF calls "desk studies": secondary analyses focusing on rather broad general topics like "worldwide migration," "international migration," and demography; around one-third of the total output can be attributed to this publication type. A second type of publications consists of empiric studies usually in the area of integration; a clear majority of 19 out of 23 empiric studies deal with integration topics.[22] Among these, there is a distinct focus on methods of quantitative sociology. In contrast to this standard design, only a small minority of five publications apply qualitative methods of social research. Out of these studies, two were designed as "supplement study" to previous quantitative surveys.[23] Furthermore, quantitative research focusses heavily on one particular source of data, namely the *Ausländerzentralregister* (Central Register of Foreign Nationals, *AZR*). In 16 projects, the raw data for drawing a sample stems from this source.

To sum it up, the analysis has highlighted some general features of the "official canon" of governmental knowledge production at the BAMF. The pub-

22 Own survey. Base: 26 research reports and 65 working papers from 2005-2015 according to the 2015 publication list (Bundesamt für Migration und Flüchtlinge 2015c)

23 Bundesamt für Migration und Flüchtlinge 2011b and Bundesamt für Migration und Flüchtlinge 2012b

lications have been analyzed in terms of their features such as topic, length, methodology, and compared against academic research. The body of official literature can be structured into two main types of research projects: first, elaborate social research projects predominantly concerned with integration and to a lesser degree labor migration, and second, "desk studies" in the areas of demography and international/worldwide migration. In regard to the object of study, two main types of research designs can be discerned. On the one hand, there are texts that can be qualified as a general overview of a social phenomenon (media use of migrants, for example[24]), or a specific area of policy steering ("The organization of Asylum procedures in Germany"[25]). These studies can usually be attributed to one of the Research Group's topical areas. On the other hand, there are studies which focus on a specific target group (integration course participants, Muslims, highly qualified migrants) and very often cut across the integration/migration divide in their analysis. By and large, the research agenda seems to reflect the main streams of academic research on migration, above all visible in the research on socio-economic aspects of integration as well as statistical overviews over selected target groups and social phenomena. After all, this approximation reflects both the academic training of the BAMF researchers as well as their strategy to conduct knowledge production using scientific methods, albeit under different production conditions.

## Qualitative Analysis

For the qualitative analysis of knowledge production, it is necessary to link the insights from the qualitative analysis to the structural conditions of knowledge production as laid out in Chapters 2 and 3. These conditions can be subsumed under the notion of practical relevance, which is understood as the dominant concept of the BAMF's knowledge production.

Practical relevance can be conceptualized as a specific practice of knowledge production developed at the BAMF Research Group as a result of both a tradition of governmental knowledge production on migration and of the structural conditions of the political-institutional set-up of governmental research at the BAMF. As the discussion of this has shown, practical relevance –

24 Bundesamt für Migration und Flüchtlinge 2010a

25 Bundesamt für Migration und Flüchtlinge 2012c

for example expressed in the strategy of acquiring study commissions, a practice which gained increasing significance over the years – is portrayed as the main defining feature of knowledge production and is seen as a core difference to academic research at the same time. Therefore, the notion of practical relevance will constitute the core item of analysis of BAMF publications.

In the analysis, the notion will be scrutinized in two ways. First, it describes the intended use of the knowledge. This follows from a notion that practical relevance is not a uniform yardstick of quality, but refers to potentially very different uses of knowledge in the context of policy-making. Second, the effects of the principle of practical relevance will be scrutinized in regard to its influence on the knowledge. Again, despite the rather starkly pronounced demarcation from academic knowledge production, it stands to reason that academic theory and method are systematically applied to BAMF-publications, albeit under the specific conditions of knowledge production present in the Federal Office.[26]

In short, practical relevance will be scrutinized both in respects to the perceived multiple uses of knowledge for governmental purposes, as well as in respect to its feedback on the structure and features of the knowledge.

These deliberations will be linked to the insights of the quantitative overview to select key areas for further inquiry. As a basic unit of analysis, the BAMF's longest or otherwise most significant research projects will be used. While this selection is not representative, it does cover a sufficiently large area of the BAMF's research activities. Based on these projects and the intended use of knowledge, four different practices of knowledge production can be discerned: administrative knowledge, depoliticizing knowledge, defensive knowledge and legitimatory knowledge. The two items – research interest and practical relevance – will be analyzed in terms of their interaction: for which exact practice are the particular studies conducted? How is academic theory selected, applied, and altered according to practical relevance deliberations? How does the theoretical understanding of a concept change over time? Before discussing the strengths and weaknesses of this approach, the four complexes of knowledge production and political use will be described briefly:

The first complex under scrutiny is knowledge for administrative purposes. This type of knowledge and practical use can be regarded in some respects as the archetypical genre of governmental knowledge: statistical re-

26 Cp. Amir-Moazami 2018b, p. 111

ports, usually in yearly intervals, complemented by legal definitions of different target populations. As demonstrated in Chapter 2, such reports constitute an important cornerstone of governmental knowledge production, which is why it is hardly surprising that the BAMF's research efforts have aimed in that direction from the very beginning. This practice of knowledge production will be discussed using the example of Migration Reports, a yearly report series issued by the BAMF since 2005.

Second, depoliticizing knowledge in the context of integration research will be analyzed. In the BAMF, the main policy area of subject formation is integration, since in this policy area the BAMF retains quite widespread authority and has developed policy tools which directly interact with migrants, such as integration courses. The basic concept of integration has been heavily influenced by academic research of the BAMF and elsewhere. In this context, two research projects are of major interest: first, the integration report as an early attempt to develop a coherent theory of integration and integration monitoring, and second, the integration panel, a long-term comparative study on integration course participants.

The third – defensive knowledge – is the use of knowledge as a remedy to populist, alarmist, xenophobic, or otherwise undesirable public statements, a relatively well-established concept of knowledge use in migration policy contexts. For the analysis of defensive knowledge, studies on Muslims will be analyzed, among these the study Muslim Life in Germany, the single most widely cited BAMF research report.[27]

Fourth is legitimatory knowledge. This type of knowledge is set apart in a way from other knowledge forms since it is directed towards a more diffuse target and therefore hardly stands strict criteria of practical applicability. In this context, the concept of migration potential and its development from a prognosis tool to a rather self-referential legitimization discourse will be analyzed. In the latter context, migration potential is adapted to interpret migration movements according to political guidelines: some migration potentials are characterized as problematic despite the fact that these hardly translate into actual migration movements (especially from Africa); in other contexts, migration potential is downplayed (intra-EU migration movements from Romania and Bulgaria). All in all, it seems probable that the practical use of these studies lies rather in the legitimization of policy than in its information.

27 Bundesamt für Migration und Flüchtlinge 2009c

Before the four types of knowledge/practical use complexes are analyzed, two caveats of this framework of analysis should be mentioned. First, as already mentioned, the four types are not an exact representation of the whole body of knowledge produced by the BAMF: some of the BAMF's research topics, such as labor market, irregular migration, or demography are not equally well-represented in the analysis as others. Furthermore, as mentioned earlier, some publication types are left out as well, most importantly the rather academically-oriented publication series, but also most studies conducted within the framework of the European Migration Network (EMN). The second caveat concerns the connection of practical use and knowledge: the construction of a typical, knowledge-practical use complex might lead to the false impression that these complexes are stable over time and somewhat clearly separated from each other. As the analysis will show, this is not the case: practical applicability requirements change over time. Furthermore, in most cases, more than one practical use of a given research project can be discerned, so it is not the case that, for example, studies on Muslims exclusively serve to calm the public debate, or the Migration Reports exclusively serve administrative purposes. Instead of understanding the typology of knowledge-power-complexes as an all-encompassing category system with fixed borders, it should rather be seen as a system of ideal types: the four knowledge-policy complexes illustrate most clearly the features of governmental knowledge production, its effects, and blind spots. This has been done by focusing on research outlets and topics which represent most clearly the "official body of knowledge" of the BAMF, which at the same time are to the highest degree subject to the institutional confines of knowledge production as illustrated in Chapter 3.

The absence of clearly confined categories can be regarded a strength rather than a weakness of this approach: since the analysis centers on the dynamic interaction of knowledge production and practical relevance requirements, changes in what is considered practically relevant knowledge can be made visible. Especially by focusing on single, long-term research projects, the evolution of theoretical concepts according to practical applicability considerations can be made visible which otherwise would stay unrevealed.

## Knowledge for Administration

Administrative knowledge production can be regarded a paradigmatic genre of governmental research because it follows most closely the instrumental logic of knowledge application to administrative action. This logic can be traced back to the concept of rational government in Max Weber's typology.[28] According to Weber, the distinctive difference between rational and other government types is the precise and planned application of knowledge to general rules within a bureaucracy.[29] Given this structural proximity between knowledge production and administration and the fact that Weber's ideal is mirrored in the Research Group's legal mandate, the original research hypothesis was that the BAMF's knowledge production is to a large degree dedicated to research questions directly arising from administrative practice. Indeed, a large part of the research publications are connected to administrative action. The most representative of these descriptive, technical publications issued by the BAMF is the Migration Report (*Migrationsbericht*), a yearly statistical overview of migration movements in Germany.

### The Migration Report

When analyzing Research Group publications, the Migration Reports do not seem to be a very promising source material: a large part of these texts is made up of either tables with statistical data or descriptions of institutional or legal regulations with no commentary or analysis. Successive reports usually update the data columns, repeat legal prescriptions, and briefly describe legal changes but otherwise contain no new information. Furthermore, the Migration Reports do not consume large resources in terms of personnel and workload: for years, the reports were compiled by the same author; until 2014, the compilation of the report was conducted in a non-scientific statistical unit.[30] The academic literature similarly seems to attribute little significance to these reports: Boswell for example judges that these rather foundational and technical publications serve predominantly a legitimizing purpose, that

28 Weber distinguishes between rational, charismatic, and traditional forms of government. Cp. Weber 1994, 311f.

29 Boswell 2008, p. 471

30 Cp. for example Bundesamt für Migration und Flüchtlinge 2011f, p. 12

the Ministry can demonstrate it has access to this sort of knowledge without actually using it in political practice.[31]

The large amount of exact repetitions of statements in the Migration Reports is indeed striking. Not only the basic structure of reporting never changes, but also the text structure within chapters is usually an exact copy from last year's report. Definitions and legal texts are repeated year after year; for example, the discussion of advantages and disadvantages of specific statistical sources stays by and large identical over the course of ten years. In this regard, these reports constitute something like the least academic publication outlet of the BAMF Research Group: there is hardly anything more punishable in the academic publication tradition than plagiarism; in any university, such a publication strategy would lead to negative repercussions.

However, contrary to the assumption that administrative knowledge is of somewhat minor interest, one of the interviewees identified the Migration Report as one of the decisive reasons for the establishment of a positive reputation of the Research Group's knowledge production.[32] Indeed, the BAMF considers the reports as one of their "standard publications" with a comparatively high degree of dissemination and political impact.[33] Given the rather negative evaluation in academic literature and visual impression of redundancy, how can this political impact be explained?

In this context, Iver Naumann's ethnography on the Norwegian Ministry of Foreign Affairs is illustrative as he discusses the production of boring speeches, specifically an incident where a relatively innovative speech to senior diplomats was drafted and subsequently rejected by a supervisor. The draft was replaced by a generic talk on the basic principles of foreign policy, a topic surely already known by this audience of experts. Naumann concludes that the most important principle of bureaucratic knowledge production is not the transmission of new information, but to maintain the basic principles of policy by way of repetition. In this context, repetition must not be seen as a lack of innovativeness, but rather, a decision for continuity. Repetition means order is upheld, since if a policy is not repeated, it would presumably be weakened.

Applied to the Migration Report, Naumann's approach is useful since it conceptualizes repetition not as a deficit, but rather a conscious strategy to

31 Boswell 2015, pp. 26–27

32 Field notes, July 2016

33 Bundesamt für Migration und Flüchtlinge 2010b, 8f.

establish a certain discourse, which can in turn be analyzed. Physically, this discourse emerges as a co-creation between involved units of a given bureaucratic organization as well as editing and the revision of texts on the various steps of the bureaucratic hierarchy. In the case of the BAMF, a similar procedure is in place which controls the public output of the Research Group though a process of hierarchical supervision and editing. The manuscript is passed on through various stages of a hierarchy, first internally (in the BAMF) and second externally (in the Ministry) where the text is scrutinized for potential political controversies. Although this is a common practice applied to all BAMF publications, it stands to reason that the Migration Reports are subject to an especially tight revision process: since the Minister of the Interior is presenting the report to the Federal Cabinet and to the press, and in general because of their relatively high public visibility, the reports are arguably controlled more closely for politically sensitive content.[34] Iterations of reporting structure which sometimes go as far as exact repetitions of whole text passages are an outcome of this hierarchical surveillance: probed and proofed concepts and formulations are used over and over again to avoid negative feedback from supervisors; the potential for innovations is systematically reduced as a result because changes are subject to especially tight scrutiny and must be justified explicitly.

Besides this insight into the production of repetitions in administrative knowledge, Naumann highlights the fact that repetitions can be, in fact, productive. Indeed, in this sense, the lack of innovation turns out to be the greatest advantage of the Migration Reports if the criteria for scientific innovation are disregarded: it is the longest established report series in the policy area of migration and enjoys relatively high public visibility compared to most other BAMF publications.[35]

In the following paragraphs, administrative knowledge production in the Research Group will be analyzed using the Migration Reports as an example. This is conducted in three steps. First, the reports are described, especially in respect to their development over the last ten years. The central focus lies on

34 Cp. press information concerning the Migration Reports: Bundesministerium des Innern 1/6/2016, Bundesministerium des Innern 12/4/2015, Bundesministerium des Innern 1/15/2014.

35 Every Migration Report is presented by the Minister of Interior to the Federal Cabinet. Cp. Bundesministerium des Innern 12/4/2015

the construction and evolution of the concept of migrant groups which constitutes a specific technique of reading statistical data. In the second step, the political usefulness of the knowledge is discussed. In this part, I follow Weber's distinction between administrative and technical knowledge, arguing that Migration Reports contain mostly technical knowledge whose central concept changes over time from the construction of an overview to a selective representation of migrant groups. In the third part, the epistemic features of this knowledge are discussed. By focusing on the knowledge on illegal migrants, this analysis challenges the notion of a somewhat neutral overview. Instead, a specific governmental perspective is created from the impression that migration is thoroughly structured and steered by legal status groups.

To draw an accurate image of the original concept of the Migration Report and its development, it is necessary to briefly outline the history of this research project. Here, the Independent Commission's critique of statistical reporting on migration can be used as a starting point, since it illustrates well the state of the art of statistical reporting of migration by the end of the 1990s:

> "In the course of its work, the commission has time and again reached the borders of transparency. A main obstacle in this regard is the fact that quality issues in migration statistics hinder an unequivocal evaluation of the overall immigration movements. The existing database is sufficient to recognize a need for action and propose political changes, but there remains a more or less large area of uncertainty."[36]

The statistical infrastructure of monitoring migration consisted at the time of the Independent Commission's report of three main sources: the Federal Statistical Office's migration statistics (*Amtliche Wanderungsstatistik*), the Central Register for Foreign Nationals (*Ausländerzentralregister, AZR*), and the Micro Census. Out of these, the most comprehensive statistical source is the official migration statistics based on population registry data. It contains information on the number, age, gender, and nationality of international migrants. While these statistics contain data on all internationally moving persons regardless of citizenship, AZR data is confined to non-Germans only. The AZR is a central database with data from foreigner authorities, the BAMF, police and other authorities where personal data on foreigners is stored. In addition to the information provided in the migration statistics data on the legal status, it contains the duration of stay and in some cases socio-economic data

36 Unabhängige Kommission "Zuwanderung" 2001, p. 287

(education, employment status) as well. Finally, the Micro Census – a panel study organized by the Federal Statistical Office based on a random survey of 1% of German households – contains a wide array of socio-economic data.

The Independent Commission's critique targeted above all else the principle of nationality: all official statistics allowed only the distinction of citizenship: German or non-German. However, during the 1990s, this distinction became increasingly less meaningful especially because of the increased number of immigrating Ethnic Germans as well an increasing naturalization rate. As a result, the binary logic of citizenship no longer corresponded with what was perceived as a social reality, as a government statistician explained.[37] Gradually, these processes rendered the hitherto quasi impermeable border between German and non-German populations more porous and challenged the image of an ethnically homogenous German population as implied by governmental statistics. Furthermore, different statistical indicators were scattered around several databases which rendered them less meaningful: data on in and out movements were compiled by the Federal Statistical Office, while statistics on legal status groups of migrants were collected by various other authorities such as the Federal Office for the Recognition of Foreign Refugees (asylum statistics), the Foreign Office (data on family reunification based on visa applications), and the Central Work Placement Agency (data on seasonal workers). With the exception of the Micro Census, all governmental statistics on migration had in common their compilation as a by-product of administration.[38] The method of data collection is in principle prone to producing bias: for example, immigration and emigration were registered in the population registries on the basis of in- and out movements, not individuals; pendular migrants therefore emerged multiple times in the statistics. As a result, definitions of what counted as migration and what not differed, so that comparability of indicators across databases was reduced.[39] A related challenge consisted of the underreporting of emigration: de-registration from official records was de jure mandatory but in practice difficult to enforce with the existing administrative and legal framework.[40] Together with the double-registration of pendular migrants, this resulted in a systematic overrepresenta-

37 Field notes, interview, February 2017

38 Lederer 2004, 102ff.

39 For example in the Central Register of Foreign Nationals (*Ausländerzentralregister*). Cp. Bundesamt für Migration und Flüchtlinge 2008c, p. 13

40 Cp. Beauftragte der Bundesregierung für Ausländerfragen 1997, p. 171

tion of migrants in the population registries, especially pendular migrants and male individuals (due to their higher degree of cross-border mobility). Another problem of registration concerned migration forms which did not correspond to legal statuses, such as work migration: data on these phenomena had to be compiled from various, partly incompatible sources with inconsistent definitions.

All in all, the critique offered by the Independent Commission targeted the patchy data basis for migration policy-making by listing the main methodological problems, such as incompatibility of data sources and definitions, inapt data collection mechanisms, and most of all, the outdated nationality principle. According to the Independent Commission, these problems cumulated to a degree that they constituted the main obstacle to policy reform.

In this context, the Migration Reports can be read as an answer to the Independent Commission's critique. By and large, the reports compile the different statistical sources on migration into one document, thereby creating a synoptic overview. The structure is made up of four parts: an overview of the total flows of migrants in a given year, a detailed discussion of selected migrant groups, emigration, illegal or irregular migration, and data on the foreign population in Germany. All these chapters are based on the main databases on foreigners in Germany. The first chapter, an overview of in- and out movements, is based on the *Amtliche Wanderungsstatistik* (official migration statistic) as provided by the Federal Statistical Office. Micro Census data is used in the "data on foreign population" chapter; the core item of analysis is dedicated to the discussion of migrant groups on the basis of AZR data. This part contains an overview of immigration movements, analyzing the following immigrant groups according to legal status and the aim of immigration: migrants from within the EU, "Ethnic Germans," temporary work migrants, students, asylum, and family reunification.[41] Over time, new groups are created – especially the "temporary work migration" and the "asylum" categories becomes more differentiated, reflecting legal changes in these two areas.[42]

41 Bundesamt für Migration und Flüchtlinge 2007a, 36ff.. Some migrant groups are registered in specialized data bases, such as different forms of work migration (at the Central Placement Agency) or data on family reunification (as part of the Foreign Office's visa statistics).

42 For example, the Migration Report 2015 names 10 forms of temporary work migration (Academic professions, managers and specialists, international staff exchange vocational training, highly qualified workers, "Blue Card EU" holders, scientists, and self-employed migrants.

In general, the discussion of migrant groups is the only part of the Migration Report which is subject to some changes in structure and a gradual expansion in volume, largely thanks to the fact that discussion becomes more detailed and refined over the years.

These changes notwithstanding, the basic understanding of migrant groups has remained stable since 2006. The Migration Reports proceed with describing every migrant group in a relatively standardized pattern: first, legal prescriptions are quoted to serve as a basic definition of every migrant group and constitute the introductory passage to every sub-chapter. Subsequently, the volume of inflow is reported, regarding the last ten years and with special attention to changes in regard to last year's report. Usually, nationality is discussed next: the most important countries of origin are listed according to their importance in terms of volume, again with a special focus on changes as compared to last year's report. In some cases, additional demographic data is presented: this includes the age and gender composition of a given migrant group and, in some cases, the regional distribution within Germany as well. All in all, the legal description of a given migrant group and its volume of inflow is the most important and consistently provided information, whereas demographic statistics are discussed only in some cases, depending on the availability of data. Socio-economic data – employment rate, income, education status, housing situation, family structure and size – indeed, any information which goes beyond the mere description of a legal status plus what can be called "passport information" (age, nationality, gender) – is consistently absent from the discussion of immigration groups.

So far, the concept of migrant groups to analyze and describe migration seems like a rather straight-forward, self-explanatory method of reading statistical data. Indeed, grouping statistical data into categories is of course not an exclusive innovation of the BAMF, but rather can be regarded as a relatively typical method of quantitative sociology of migration. However, there is an important difference to earlier concepts which used a similar heuristic: in the 1970s and 1980s, research and policy-making centered on certain "risk groups" such as Second Generation, Turkish Nationals, or Asylum Seekers. These risk groups were specifically selected according to a perception of danger or being endangered – in short, of being in need of governmental intervention. Another related concept consists of national groups in the tradition of "Guest Worker" research which are usually used to compare the largest foreign populations to each other.[43] These include most often the following "Guest Worker" nations: Turkey, Former Yugoslavia, Italy, Greece, and Spain. In contrast to these earlier concepts, the Migration Reports attempt to cover *all* migration movements through the lens of migrant groups based on their legal status. At first glance, the logic of legal status is compelling: it avoids the problem of national groups which become increasingly harder to separate from each other in a meaningful way. This can be be illustrated with the question who would count, for example, as an "Italian" in the classic "Guest-Worker" nationality comparison scheme: The traditional concept of citizenship became less meaningful as an indicator of social and economic status after the rate of naturalizations grew, so that not only Italian passport holders, but also their naturalized spouses and offspring could be added to this national category. The problem multiplies when including the second and third generation of migrants: are only those with "pure" Italian ancestry regarded as Italians, that is, two Italian parents and four Italian grandparents? Or would some German passport holders in the ancestry line be accepted, and if so, how many? What about second generation immigrants with a bi-national family background, say a Turkish father and an Italian mother? Even if these questions could be answered, Data collection would be the next large problem: Methodologically sound data would have to consist no less than 16 sub-groups to account for all the possible combinations of German-Italian ancestry until the second generation; drawing a sample with reasonably large populations in all of these sub-groups would be dauntingly difficult and costly. Even if this data could be collected, and not only for Italian, but also for the other large immigrant groups, its practical use seems to be questionable. As already mentioned, the five largest foreign national groups cover a decreasing share of all migrants, which is why the concept of focusing on the largest national groups became more and more outdated over the last years. In contrast to this approach, the BAMF's migrant groups offer the advantage that they are clearly defined by the legal regulations governing them. This logic is based on both a methodological and a legal aspect: not only does it follow the technique to steer migration through the distribution of rights and handicaps through legal titles, but also from a pragmatic consideration that official data is usually structured according to these legal definitions. This data adds qualitative information to the two most commonly provided statistics in migration reporting, making it

43 Cp. for example Bundesamt für Migration und Flüchtlinge 2008g

is possible to report not only on the volume and nationality of migrants, but also on their purpose of stay.[44]

In relation to the analysis of knowledge production conditions at the BAMF, some factors in the institutional make-up of the Research Group can be identified which influenced the shaping of the concept of migrant groups. The groundwork goes back to research at the European Forum for Migration Studies (*Europäisches Forum für Migrationsstudien*, efms); a research institute associated with the University of Bamberg.[45] One of the researchers, Harald Lederer, developed the concept of migrant groups which could eventually develop into "alternative foreigner migration statistics" on the basis of AZR data.[46] Lederer was later one of the first members of the Research Group and became the responsible author for the Migration Reports for years to come. Another reason for the development of the migration groups as an alternative to the immigration statistics is the BAMF's exclusive access to AZR data and a corresponding interest to use this data source as a strategic resource. In the use and discussion of AZR data, the structural conditions of coordination and competition between the various providers of governmental knowledge become apparent. This means that on the one hand, data from other state agencies is duly quoted and discussed, which reflects the effort to include all relevant providers of knowledge into the Migration Report. On the other hand, the BAMF frequently underlines the superior quality of AZR data, especially in comparison with the Federal Statistical Office's immigration statistics:

> "Immigration statistics as a basis for integration policy-making have to provide quantitative data for the single immigration groups which feature different preconditions for their stay in Germany. These differences originate in the different legal frameworks which govern the immigration and stay of these groups. Since the German migration statistics do not provide information on the purpose of stay, the Migration Report differentiates between distinct migration forms."[47]

To sum it up, in the initial phase of research at the BAMF, the provision of general knowledge was not confined to the Migration Reports but constituted

44 Lederer 2004, 44ff.

45 See Bade 2017, 63f. for detailed description of the EFMS' institutional structure.

46 Lederer 2004, 69 and 248ff.

47 Bundesamt für Migration und Flüchtlinge 2006, p. 11

rather something like a general principle of knowledge production: many research projects from the Research Group's initial phase can be characterized as an "overview," "collection of data," or "basic information."[48] The Migration Reports are in this respect a paradigmatic report series, since they contain a comprehensive collection of official data. In this context, the concept of migrant groups can be regarded as the key item for interpreting this data. Frequently, the reports state that this concept allows for a more thorough, detailed discussion of migration, which in the end leads to an increased practical relevance for policy-making.[49] In the concept of migrant groups, the generation of such knowledge can be studied in detail: on the one hand, migrant groups answer the critique put forward by the Independent Commission with technical improvements, such as merging different statistical data in one report.; on the other hand, these improvements extend to the conceptual area as well, since migrant groups offer a simplified overview of the migration movements. However, this approach stands in competition with other concepts, most importantly Migrant Background. Migrant groups are in fact a partial view of the BAMF on the phenomenon of migration, shaped by specific institutional effects. In the implementation of the migrant group concept, two of these effects have been described: first, personal continuity from efms staff to the Research Group and a consequential transfer of knowledge; and second, the increasing reliance on AZR data as a result of exclusive access rights and a corresponding interest to promote this data on the side of the BAMF.

## Practical Relevance: Legibility

As already mentioned, the Migration Reports constitute in some respects the "least academic" publication form of the Research Group, especially considering the high degree of technical information, repetitions, and copy-pasted passages of earlier editions. Despite the technical character of these publications, the question of practical applicability is not as easy to answer as one might expect: it stands to reason that in general, a large part of administratively relevant knowledge is produced and distributed in less formalized

48 According to a former Research Group member, between 1/3 and 2/5 of early research projects (published before 2011) were self-commissioned. Among later publications (from 2013 on), self-commissioned projects are almost completely absent.

49 Eg. 32 pages in 2004, 111 pages in 2009, ca. 80 pages is 2013. Cp. also Bundesamt für Migration und Flüchtlinge 2009b, p. 16

ways, such as files, memos, submissions, and tacit forms of knowledge and bureaucratic practices.[50] In the literature, this type of knowledge is usually referred to as "administrative knowledge" following Max Weber's notion of *Dienstwissen*.[51] Administrative knowledge is characterized by the fact that it is generated as a by-product of administration and is usually not publicly available;[52] both criteria do not fit well with the production conditions of the Research Group, as empirical evidence from interviews seems to confirm. As an example, according to an official of the Ministry of the Interior, most parliamentary inquiries (*Kleine Anfragen*) are usually passed down the hierarchy and answered by government officials on the basis of past experiences with similar cases; academic knowledge production is usually too time-consuming for official memos, ministerial requests and other non-public forms of administrative knowledge.[53] Bearing this in mind, the actual usefulness of the Migration Reports seems less straight forward to pin down – after all, the Migration Reports constitute something like the BAMF's flagship publication. In the next paragraphs, the practical applicability of administrative knowledge will be discussed.

By answering the Independent Commission Integration's critique of inapt statistical reporting, a window of opportunity for the Research Group to provide politically relevant knowledge presented itself. With the concept of migrant groups, the BAMF was keen to introduce a reformed perspective which would establish something like an overview of migration statistics. It seems clear that the initial concept of the Migration Report exceeds the use of legitimatory knowledge, since the Independent Commission, the Federal Parliament, and other political actors clearly articulated demand for this type of administrative knowledge. The BAMF invested some resources into this project since two university researchers were specifically recruited for this task.[54] Clearly, the Research Group as well as the BAMF leadership expected some political impact from this type of research.

---

50 Mangset and Asdal 2018, p. 2

51 Weber 1994, p. 373

52 Quoted after Affolter 2017, 145f.

53 Requests are a form of relatively widespread, low key parliamentary action which are submitted by an opposition-party parliamentarian. The request is answered by officials of the respective ministry, or, if applicable, by BAMF staff. Field notes, September 2013

54 Bade 2017, 63f.

Coming back to the question of political usefulness, This type of foundational research points to a specific practice of governmentality which can be described in James Scott's notion of legibility: in his study "Seeing like a State," Scott described several practices of knowledge production and governance which he characterizes as constitutive for governance:

> "I began to see legibility as a central problem in statecraft. The premodern state was, in many crucial respects, partially blind; it knew precious little about its subjects, their wealth, their landholdings and yields, their location, their very identity. It lacked anything like a detailed 'map' of its terrain and its people. It lacked, for the most part, a measure, a metric, that would allow it to translate what it knew into a common standard measure necessary for a synoptic view."[55]

Scott refers here to practices of governmental knowledge production in the early modern era such as mapping of forests, the introduction of citizen registries, or the creation of a cadastral map. In all these cases, the collection of data, the creation of an overview, is not only motivated by a scientific-forensic view to map and understand a complex phenomenon, but ultimately, by the political requirement to control and steer. The collection of data is usually organized in a way that the phenomenon under scrutiny becomes readable from a distance, thus prepared for political intervention. In other words, as Zachary Karabell puts it, "what gets measured gets managed."[56] Scott's notion of legibility can therefore be regarded as a specific form of practical use of basic statistics: by establishing a synoptic view, a specific order of a complex phenomenon is created which can be seen as a necessary precondition for political steering. In some respects, the Migration Reports can be read like an attempt to make migration policy readable to the policy maker: migrant groups introduce a qualitative order into the chaotic stream of immigration from an extremely diverse background, governed by multiple legal schemes both on the EU and national level and captured in multiple, partly contradicting data sources. By broadly sorting these streams into larger categories (humanitarian, temporary work, family reunification, Ethnic Germans), a specific logic of knowledge is introduced which makes the phenomenon readable and understandable. The legal groups, together with the non-category of illegalized

55 Scott 1998, 17f.

56 Karabell 2014, p. 13

migrants encompass the sum of what the state can know about migration and therefore represent something like a "synoptic view" of the phenomenon.

However, while the first Migration Reports look like a promising answer to critique, they developed in a different way than expected if the criteria of a "legibility" practice of knowledge production are used. This manifests in two ways: the internal division of the governmental perspective and the lack of a political impact by the Migration Reports.

Concerning the first point, it is important to note that almost at the same time as the concept of migrant groups emerged, a very similar concept was put forward by the Federal Statistical Office on the basis of Micro Census data called Migrant Background. In 2005, not only the citizenship but also the country of birth of a respondent and their parents became part of the Micro Census questionnaire, thus allowing for a distinction between migrants, their first and second generation offspring, and naturalized citizens. These population groups were subsumed under the category "Population with Migrant Background." This new concept can be regarded as an attempt to overcome the outdated German-foreigner divide in the migration statistics. Migrant Background is a quasi-ethnical category which avoids overtly ethnic categorization, but allows for tracking socio-economic features of migrants and their descendants regardless of citizenship.[57]

Concerning the lack of political impact of the BAMF's knowledge, it is important to keep in mind that Scott's understanding of legibility entails a logic of both systematization and political intervention. While the Migration Report and especially the concept of migrant groups can be seen as an attempt to systematize migration reporting, the second part of Scott's concept is somewhat lacking. This is not a coincidence, but rather can be regarded as a consequence of the legislation process of the Residence Act, as described in Chapter 3.2., where the technocratic principle of a scientifically-steered migration policy was systematically diminished to retain political control over the area of legislation. While the original concept of immigration policy-making included a system of recruitment of fixed quota of migrants based on expert recommendations, such a systematic relationship between political decisions and statistical reporting is absent in the case of Migration Reports. EU migration, the largest single source of migrants, is to a large degree independent of political steering; other migration streams, such as family reunification and asylum, are largely dependent on factors which are beyond political control

57 Salentin 2014

as well. In the literature, European statistical reporting projects are discussed as a similar case in this respect: Eurostat started to compile statistical data on migration-related topics such as asylum and irregular migration around 2005 to provide facts about these policy areas, but the political impact was likewise marginal.[58]

In sum, these findings point to the fact that the provision of statistical data alone does not guarantee its political relevance. In consequence, the construction of a synoptic view loses most of its political significance if there is no corresponding possibility or motivation for political intervention.

This observation leads to the question of how the Migration Reports have developed over recent years. If the legibility concept is taken seriously, it should lead to a systematic expansion of knowledge production and the eradication of incompatibilities in the statistical reporting. However, this is not the case: in principle, almost all blind spots, flaws in the statistics, incompatibilities and so on which were documented in 2005 are still in place ten years later. This leads to the peculiar situation that single migrant groups cannot simply be added up to calculate a total number of migrants, a major drawback of the concept as compared to the official migration statistics and the Micro Census.[59] While this is duly discussed and protocolled, no actual progress has been made in this regard since 2005.

While the reasons for this lack of development are not explicated, they could lie in the inherent problems of the cognitive frame of migrant groups: from the beginning, the concept tried to unify partly incompatible statistical sources, as already mentioned. Furthermore, the AZR as the main data source for the migrant groups is not without its problems in terms of data quality. First, the database contains only data on foreigners, which excludes by definition all naturalized citizens. This fact is mentioned in the Migration Reports, although its potential for introducing bias especially in regard to integration monitoring is not: the AZR's basic logic of foreign nationality risks excluding the most successful or otherwise privileged migrants – those with access to citizenship – from the basic sample. Second, the nature of the AZR as a primarily administrative database could further aggravate this bias potential: as AZR data is automatically updated by the authorities in the course

58 Kraler et al. 2015, p. 46

59 Bundesministerium des Innern and Bundesamt für Migration und Flüchtlinge 2016, p. 52

of administrative acts, foreigners with a relatively unsafe legal title are overrepresented in the database since they are legally required to interact with authorities more often. Additionally, the most privileged immigrant group, EU nationals, are rarely registered in the AZR due to data protection restrictions. In total, AZR data overrepresents holders of short-term or otherwise precarious legal titles, excluding at the same time naturalized persons and EU citizens, which is, in the end, an approach with a built-in bias towards overrepresenting the "least successful" in socio-economic terms.[60] When this data is used to construct migrant groups, it has a tendency to hide the connection between socio-economic integration and the distribution of legal privileges. This follows from the fact that only legal status information without the corresponding socio-economic data is stored in the AZR. If, for example, the criminal rate among refugees or the unemployment rate among short-term status holders is measured, it is unclear if these attributes are caused by the lack of legal security or constitute rather an inherent characteristic of the respective group. An in-depth longitudinal analysis concept could trace migrant status careers and link them to socio-economic data, thus highlighting the effects of privileges and restrictions connected to certain status groups.[61] Since this sort of analysis is impossible with AZR data, differences between groups appear as if they were exclusively caused by inherent individual characteristics rather than by unequal legal prerequisites.[62] These drawbacks confine the AZR data to a relatively narrow area of reporting of immigration of third country nationals[63] at the moment of their immigration.

Even if these methodological problems were solved, another problem appeared which concerns the construction of a category system of immigrant groups. In 2005, when the categories were first designed, immigration groups typically held between 20,000 and 50,000 persons each – with the exception

60 Salentin 2014, p. 25

61 In a recent research project (BAMF-SOEP-IAB Panel on refugees), a similar research design has been set up to monitor the long-term societal integration of refugees, for the first time including data both on legal status and socio-economic integration. Cp. Bundesamt für Migration und Flüchtlinge 2016a

62 Bundesamt für Migration und Flüchtlinge 2008e, 38f.

63 In the terminology of the BAMF, "third country nationals" are citizens of non-EU countries.

of EU migrants, whose number amounted to ca. 280,000 persons.[64] By and large, the migrant groups were numerically in a similar order of magnitude and therefore comparable to each other; this fact led to the concept of an intuitive plausibility. Meanwhile, this relative balance has shifted starkly: "Ethnic German" migration continually lost significance in terms of volume, dropping from ca. 35,000 persons in 2001 to merely 6,000 in 2015.[65] The category of work migration similarly decreased, especially due to a reduction in seasonal migrants. At the same time, with the introduction of additional legal instruments especially for highly-skilled and highly-qualified migrants, the category of "temporary work migrants" became more and more diversified internally;[66] what started out as essentially a legal status for low and semi-skilled workers gradually grew into a multitude of temporary work titles for both low skill and highly-qualified workers. The Migration Report studiously list a multitude of national and EU visa programs for specialists, entrepreneurs, researchers and other highly-qualified work migrants but with a volume of a few hundred cases each: all of these various temporary-work-related status groups make up less than 2,000 persons combined.[67] In contrast to this very detailed reporting, the most significant group in terms of volume of inflows – EU migrants – is hardly discussed at all. This migration stream gained importance, since between 2005 and 2015, immigration from within the EU more than tripled. However, this highly diversified immigration movement is not analyzed in the same depth as it is in the case of the different immigrant groups of third country nationals: neither the aim of migration, nor the duration of stay is discussed in the context of EU-migration; merely data on the most important countries of origin, as well as on the gender and age of EU-migrants, is presented.[68] This is again a result of the challenge to compare data across databases as discussed above: EU nationals are rarely registered in the AZR, so the Federal Statistical Office's immigration statistics are used. As a result, potentially very different migration projects are grouped together

64 The smallest group, Jewish immigrants, being somewhat an exception to this rule with ca. 6,000 migrants in 2005. All numbers quoted after Bundesministerium des Innern and Bundesamt für Migration und Flüchtlinge 2016, p. 52

65 Bundesministerium des Innern and Bundesamt für Migration und Flüchtlinge 2016, p. 121

66 Such as "Blue Card EU" from 2012, or work visa for scientists

67 Bundesministerium des Innern and Bundesamt für Migration und Flüchtlinge 2016, p. 52

68 Ibid., 46ff.

in this category which is analyzed only in regard to basic demographic data; the two main advantages of AZR data – purpose and duration of stay – cannot be applied to the most significant migrant group in the Migration Reports. While initially, the migrant groups establish an order through which statistical data on immigration could be made comprehensible and comparable, this order increasingly lost its plausibility over the years. Some migrant groups decreased in volume, others increased multifold. In this respect, the concept suffers from the low degree of flexibility in reporting: the system of migrant groups can become more detailed (as in the case of asylum seekers and temporary work migrants) but not more flexible; the basic order of reporting always stays the same.[69]

As a consequence, these methodological shortcomings led to an abandonment of a "synoptic view" around 2012. While the Migration Reports continue to be published, there is a subtile but important shift in the construction of migration groups. Since 2012, migrant groups do not contain all immigrations to Germany any more. This is most importantly visible in the fact that EU migration is now discussed in a separate chapter, a decision which somehow downplays the numerical significance of this migration form since the discussion is relatively brief and superficial. In the 2015 report, more than 850,000 immigration and 500,000 emigration acts are discussed in five pages only.[70] The overview perspective of the earlier Migration Reports is replaced by a detailed description of migration broadly separable into two main migration status groups: on the one hand, temporary work migrants, highly- skilled migrants and international students; on the other, asylum seekers and immigrating family members whose immigration is subject to gradually enlarged preconditions to curb these migration streams. The new epistemic order of the Migration Reports seems to reflect a new understanding of practical relevance: not the creation of an overview, but rather the criteria for political intervention vis-a-vis different migrant groups from third countries seems to be the basic logic behind the migrant groups from 2012 on.

In summary, in its original design and purpose, the Migration Reports can be considered a fairly typical project of legibility: the state attempts to understand (and ultimately control) a complex phenomenon by introducing a

---

69 Cp. Schiffauer 2018

70 Bundesministerium des Innern and Bundesamt für Migration und Flüchtlinge 2016, 46ff.

harmonic, coherent order which makes the phenomenon readable from a distance. With a focus on legal prescriptions and demographic data, the state's view on the migrant groups is thoroughly inscribed in the logic of knowledge. The role of the Migration Reports as the most widely distributed and recognized publication testifies to the success of this concept. However, Scott's concept does not accurately describe the further development of migration reporting. This becomes apparent in two points: on the one hand, the notion of a "governmental perspective" creates a false sense of uniformity of state actors; and second, there is strong evidence that the "legibility" concept has been abandoned after some time.

Regarding the first point, Scott's understanding of legibility creates a sense of a uniform perspective of the state which does not reflect the institutional competition for expertise and influence. The BAMF's concept of migrant groups stood in direct competition to the Federal Statistical Office's concept of Migrant Background. Both concepts were designed around a specific database to which the respective authority had preferential access (AZR in the BAMF's case and Micro Census data in the Statistical Office's case). Some hints of this institutional competition are subtly visible in the Migration Reports, where the advantage of AZR data is frequently emphasized which spells out to the informed reader as a competitive advantage over rival data bases. The disadvantages of the AZR, most importantly its socio-economic bias towards disadvantaged legal status groups are not equally well-documented, however. While both the BAMF and the Federal Statistical Office feed a discourse of objective reporting through statistical data, the actual practice of knowledge production is different which leads to the conclusion that in fact, it is imprecise to speak of a unified, coherent "governmental perspective."

This conclusion fits well to the observation that the governmental purpose behind this knowledge order has evolved from a "legibility" concept to a selective representation of those migration streams which are subject to political control. This evolution is mirrored in the changed concept of migrant groups: while in the first years of statistical reporting, the discussion of migrant groups was promoted as an alternative to the official migration statistics, this approach was somewhat altered after 2012, when by far the largest migrant group (EU migrants) was excluded from the respective chapter. The idea of an all-encompassing report was replaced by a more selective discussion of various migrant groups of third-country nationals; the logic behind these groups is no longer a criterion for quantitative volume but rather a criterion for the possibility of political intervention.

## Effects on the Knowledge: Governmentality

In the last paragraphs, the question of political applicability of the BAMF's production of technical, administrative knowledge has been outlined. By analyzing the evolution of the concept of migrant groups, a distinct development from a general overview of migration to a rather particular representation of selected immigration groups has been detected. In the following paragraphs, the effects of practical applicability considerations on the knowledge will be discussed, mainly focusing on bias sources arising from this particular institutional and theoretical arrangement. These bias sources entail on the one hand a tendency to internalize the perspective of the nation state defined as methodological nationalism in the literature. This is analyzed with the example of the Migration Report chapter on emigration. On the other hand, bias arises from the specific construction of migrant groups as neatly separated social units which will be demonstrated with the example of the discussion of illegal migrants. The central hypothesis is that both bias sources create a specific perspective of governmentality, in which migration is made to appear like an orderly stream of in- and outflows under administrative control.

The effect of creating a governmental perspective is well exemplified in the discussion of "illegal" migration in the Migration Reports.[71] By and large, the chapter employs the same heuristic tools as in the case of other immigrant groups. As usual, this section starts with a legal definition of irregular migration: according to the BAMF, an illegal migrant is a person without a passport, without a legal status, or someone who is not registered in the Central Registry of Foreign Nationals or elsewhere.[72] Notably, this understanding

71 Contrary to the other sections of the report, this chapter has changed its title quite often: in 2005, it started out as "uncontrolled migration"; in the next report, it was renamed to "illegal migration"; the title from the 2007 reports onwards reads as "illegal/irregular migration." In this text, the term "illegal" is used in parentheses to maintain on the one hand a consistent vocabulary which is used by the BAMF as well; on the other hand, the parentheses are due to the constructed, biased and partial representation of the term in the BAMF's understanding. Cp also Ferguson 1994

72 Bundesministerium des Innern and Bundesamt für Migration und Flüchtlinge, p. 137; cp. also Bundesamt für Migration und Flüchtlinge 2005b, p. 3

does not exactly mirror the legal definition of illegal stay, since it excludes rejected asylum seekers as well as migrants who have acquired a legal title under false pretenses.[73] The BAMF's definition follows the pragmatic understanding prevalent in the foreigner administration. An official stated in this regard that "we do not know the 'classic' illegal migrant. He ceases to be illegal as soon as we know him."[74] All in all, to the BAMF, the criterion of "unregistered" or "undocumented" migration is the crucial item of this definition; in this non-status, the definition resembles other migrant groups in the Migration Report. Next, the quantitative volume of the illegal population is discussed. Here, a principal difference to the other migrant groups becomes apparent, since this data is notoriously difficult to provide: in every volume since 2005, the Migration Reports state that there is no reliable data on the "illegal" migrant population. The reports name mainly methodological reasons for this lack: first, by definition, illegal migrants are not registered in the usual databases which are used in the other parts of the migration report, most importantly the Central Register of Foreign Nationals; second, in this context, the notion of migrant agency is discussed:

> "Both illegal entry and illegal stay are criminal acts which is why undocumented foreigners are keen to hide their stay from official registration. State authorities are in principle obliged to inform the police if they register an illegally staying foreigner [...]. All in all, illegal migrants are to a large degree hidden from statistical registration."[75]

Publicly debated estimates on the number of illegal migrants are criticized as "unsubstantiated" and "inapt for political decision-making."[76] Instead, the reports present two indicators on apprehension of illegal stayers and attempted illegal entries.[77] However, these statistics are also problematic from a methodological point of view: both databases cover only cases known to the authorities and are unrelated in any systematic way to the unknown total number of undocumented migrants. Furthermore, an analysis of trends in the data

73 Salentin 2014, p. 36

74 "Den klassischen Illegalen kenne ich nicht, und wenn ich ihn kenne, dann ist er's nicht mehr.". Quoted after Vogel and Aßner 2012, p. 24

75 Bundesministerium des Innern and Bundesamt für Migration und Flüchtlinge, p. 137

76 Ibid.

77 Bundesamt für Migration und Flüchtlinge 2006, 89ff.

is virtually impossible since the legal and political basis of these police activities is relatively instable. For example, changes in apprehensions at the borders since 2004 are much more likely caused by the EU accession of the neighboring states to the east than by actual changes in irregular migration. Basically, the Migration Reports deliver only the raw data from two relatively insignificant indicators, the critique of "unsubstantiated estimates", and the claim that the aforementioned indicators are superior despite their very limited significance. Despite the shortcomings of the approach, and despite the fact that the BAMF had issued a study with an improved methodology on the estimation of the illegal population, the basic structure of statistical reporting in the Migration Reports never changes.[78]

In sum, while the chapter on illegal migration is structured differently from other migrant groups, the basic heuristic is comparable: the definition refers to legal aspects which can then be connected to statistical databases (or in this case, to the absence of statistical reporting). While the patchy database for illegal migration is repeatedly criticized, the reports nevertheless draw an image of irregular migration which can be characterized as a criminalistic description of the phenomenon. The injury of the law is the point of departure for the argumentation; this injury is presented as a result of the individual action of a perpetrator. In its core elements, illegal migration is conceptualized in the same way as all other migrant groups: definition, statistical data, purpose of stay.

By discussing the administrative measures to combat illegal migration, and dismissing most other voices on the topic as unsubstantiated, a perspective of governmentality on the subject is created. On the basis of the notion of non-registering, the phenomenon of undocumented, irregular migration is perceived as if it was another more or less stable population group with common attributes as in the case of other migrant categories discussed here. However, by way of definition or selective analysis, many of the specific features of this phenomenon which contradict this perception are excluded: the definition of undocumented migrants is by itself incomplete since it excludes registered persons who are nevertheless obliged to emigrate, most importantly rejected asylum seekers. Illegality is therefore incorrectly portrayed as the result of a perpetrator's choice to, for whatever reason, disobey the law. Furthermore, this concept of an "illegal population" masks the fact that the

---

78 In 2011, the BAMF published an update to the methodology of reporting of irregular migration. Cp. Vogel and Aßner 2012, p. 39

borders between irregular and regular migration are not clearly defined. This can be seen by the fact that most irregular migrants enter legally and overstay their visa; at the same time, AZR data shows that regularization of formerly irregular migrants is relatively widespread.[79] With other words, status changes in both directions between legal and irregular are common. This observation contradicts the assumption of a stable illegal population which can be analytically separated from other migrant groups.[80]

Overall, the discussion of illegal migration reveals basic features of the governmental character of the knowledge produced here: on the one hand, the database is very patchy and incomplete, as the authors duly note. On the other hand, the present concept is still considered superior to every other vague and incomplete discussion of illegality, simply because it stems from official databases. This is notable since updated methods commissioned by the BAMF (which look at least superficially more promising) are ignored in this context. Instead, it seems that the superiority of governmental knowledge is not based on actual technical excellence. Rather, there is a defensive mechanism at play: no one can be blamed for using official data, even if it turns out to be biased. This cautious and conservative knowledge production can be connected to the general restraint of theory development as discussed in the context of departmental research. In the case of the Migration Reports, it is further enhanced by the comparatively high public visibility which creates potential for negative media attention, especially in connection with politically sensitive topics such as illegality.

From a political relevance point of view, the lack of innovation can in fact be seen as a useful asset of the Migration Reports, since too accurate knowledge could turn out as a political disadvantage for the Ministry of the Interior in the case of illegal migration. Take, for example, the assumption that most illegal migrants use legal documents obtained in one way or another to live a relatively undisturbed life. Although the problem seems to be sizeable, such tactics are not especially targeted by the authorities; furthermore, in the Migration Reports, these illegal migrants are by definition excluded from reporting. If, hypothetically, the BAMF reported on these tactics with a more comprehensive analysis method, such knowledge would immediately pressure the ministry to act. In her analysis of expert knowledge on illegal

79 In 2010, a consistency check of AZR data revealed that more than half of the irregular stayers had meanwhile acquired a residence title. Cp. Vogel and Aßner 2012, 25f.

80 Vogel and Aßner 2012, p. 39

employment in the UK, Boswell states that in similar cases the British Home Office is compelled to accept pragmatically knowledge gaps to avoid political trouble.[81] Somewhat ironically, this means, in the case of illegal migrants, politically useful knowledge is one that is not very precise, accurate and up-to-date to avoid negative public feedback.

It is worth to scrutinize the effects of this institutional structure as well as policy relevance on the knowledge generated. Two major sources of bias can be discerned. First, legal status groups emphasize a perspective of governmental steering which overestimate the impact of legal status management policies especially in the face of migrant action. For the documentation of migration from a political-legal perspective, this perspective introduces a governmental perspective of steering: a change in one or the other status group – say, more green card holders or less irregular migrants – can describe an actual change in migration patterns – or just be a result of tactical switching of legal statuses on the side of the migrant. Again, since statuses are not socially confined, these represent less separated, fixed populations but rather fluid categories which are perceived as opportunity structures by migrants. This is especially problematic in the discussion of irregular migration, which displays the inherent characteristic weaknesses of the category system quite well: by way of establishing irregular migration as a migrant group, the false impression of a stable illegal population group is fostered. However, illegality is characterized by a very high fluctuation of legal titles, social practices, and a comparatively patchy database. Illegality is more often than not a transitory phenomenon dependent mostly on migrant action and not on governmental steering. Even if the data was better, the category system could not cope with the inherent dynamics of the phenomenon. A category system is inherently static; it can be improved by increasing the number of categories, but their definitions will remain rigid.

The concept of migrant groups does, however, produce a specific perspective of governmentality: by focusing on migration forms that are subject to administrative control – either of an enabling nature, such as highly-skilled migration, or in a restrictive sense, such as family reunification, irregular migration, and asylum – migration appears like a top-down, steered, orderly process under the auspices of the government. In the case of illegal migration, this aspect can be seen in the extensive discussion of technical surveillance

---

81 Boswell 2011, 21f.

measures to counter the phenomenon; if the length and depth of the discussion can be used as a proxy, this aspect becomes increasingly important to the overall analysis between 2008 and 2010.[82] As the discussion of political usefulness has shown, the overall focus of the Migration Reports is increasingly influenced by the logic of governmentality. Together with the neoclassic theory of migration as an individual cost-benefit operation, this notion of governmentally-steered migration movements seems to be an increasingly influential concept:

> "Two types of belief have been particularly influential in migration policy formation. One is the economic classical theory, according to which people move to maximize their individual utility (usually through higher income) and cease to move, or return home, if the cost-benefit equation changes. The second is the bureaucratic belief that regulations designed to categorize migrants and to regulate their admission and residence effectively shape aggregate behavior. Together these two beliefs add up to the idea that migration can be turned on and off like a tap by appropriate policy settings."[83]

In the Migration Reports, another aspect of governmentality is visible when considering the institutional structure behind the knowledge: it stands to reason that not an abstract governmental logic is at work here, but rather the concrete governmental perspective as present in the BAMF and the Ministry of the Interior. The focus on migration streams that are controlled by either institution (such as asylum, or highly-skilled migration) take clear precedence over those that are not (such as EU migration). This institutional structure is similarly influential on the use and discussion of statistics: AZR statistics as the database under immediate control of the BAMF are discussed relatively positively, while data from other authorities is subject to more critical feedback. The potential drawbacks of the AZR as discussed above are to a degree omitted not because they are irrelevant, but because such a discussion would diminish the strategic usefulness of the knowledge vis-a-vis other institutions.

82 Bundesamt für Migration und Flüchtlinge 2011f, 166ff.

83 Castles 2004, p. 858

## Conclusion

To summarize, the Migration Reports – as the most important representative of a group of administrative knowledge publications – constitute a particular knowledge order initially designed mainly for the purpose of making migration as a complex phenomenon readable to the state administration. Historically, Migration Reports originated from a statistical data collection to provide an overview of all migration-related phenomena; by doing this, the Migration Reports display an unusually high degree of repetition which initially led to the false assumption that these reports were of minor interest when analyzing governmental knowledge production. However, despite the repetitions and the fact that Migration Reports almost never contain information which can be considered surprising or unforeseen, this report series is the BAMF's most widely disseminated publication.[84] Somewhat ironically, it seems that the project with the least degree of scientific innovativeness renders the greatest amount of public visibility of the BAMF research work. It seems clear that a deficit perspective – that of lacking scientific innovativeness – does not fully capture the significance of this report series. It stands to reason that this lack is not, in fact, a deficit of bureaucratic knowledge production, but rather serves an important discursive function. To quote from Naumann's concept of bureaucratic knowledge production:

> "It takes hard discursive work to keep things as they are. Making the world seem stable when it is in fact in constant flux means that wielding power involves the ability to freeze meaning. This has to be done by constantly repeating specific representations of things, actions, and identities, until what one repeats is naturalized to such an extent that it appears doxic."[85]

With orderly statistical reporting and the resulting concept of migrant groups to interpret the statistics, an image of a government-regulated policy area is drawn.

The most important theoretical innovation in the Migration Reports is the concept of migrant groups. This concept has been introduced as an answer to the Independent Commission's critique of statistical reporting, by taking advantage of the BAMF's exclusive access to AZR data. This concept is presented as a step towards more accurate and detailed statistics which overcomes the

84 Cp. for example Bundesamt für Migration und Flüchtlinge 2015a, p. 20

85 Neumann 2012, 79f.

outdated nationality principle and introduces data on the duration and aim of migration. However, over time, it turned out that the BAMF was unable to overcome the basic problems of data collection with this concept. Due to the fact that the main data source, AZR, displays a bias towards overrepresenting the least privileged migrants, and otherwise lacks socio-economic data to document integration trajectories, the initial plan of "alternative migration statistics" has been given up. Instead, since 2012, the Migration Reports focus increasingly on those migrant groups which seem important to the BAMF or the BMI. These include mostly third country nationals who immigrate either on humanitarian grounds or as temporary work visa holders. Excluded from this view are, most importantly, EU nationals which as a group constitute a large majority of all immigrants, as well as some types of illegalized migrants. In sum, what started out as a general overview in the sense of a "legibility" project turned into a very specific representation of migration as seen by the BAMF: migration is described as if it is governmentally steered with the attribution of several status titles, while the effect of these status titles on integration processes is hidden from analysis.

## Integration Research

In terms of research output, integration research is one of the most important research areas at the BAMF Research Group. Between 2005 and 2015, around one-third of the projects were conducted with this topic.[86] There are two historic reasons for the outstanding significance of the topic. First, in general, integration was identified as the core task of migration policy-making in the Independent Commission's report to make up for policy failures during the "Lost Decade". In this context, as will be discussed later in some detail, the BAMF's research can be placed in a larger discourse on governmental integration concepts which emerged around the turn of the millennium. A second reason can be found in the historical context of the institutional reconfiguration of the BAMF in 2005, where a direct line between the foundation of the Research Group and the implementation of integration policy can be drawn:

> "A number of commentators within the migration policy community have suggested that this became more or less inevitable once the agency was ascribed a new mandate for dealing with integration. With its new competence in this area, it was clear that it would need additional expertise to inform its activities."[87]

In the next paragraphs, some aspects of the knowledge production which can be characterized as general integration research will be analyzed. The central hypothesis in this chapter is that integration research developed from an initial plan to create a theory-informed understanding of integration policy-making in the form of indicator-monitored statistical reporting. While this monitoring system has been abandoned, some parts of the theory as well as the technocratic understanding of policy-making have been implemented and shape a specific governmental understanding of integration; its effects will be discussed using the example of discrimination in the context of the integration panel.[88]

---

86 31 out of 109 publications and 39 out of 125 research projects. Data according to Bundesamt für Migration und Flüchtlinge 2015c and Bundesamt für Migration und Flüchtlinge 2016b

87 Boswell 2009b, p. 164

88 The Integration Report consists of ten working papers published between 2008 and 2012 with various topics such as labor market participation (Bundesamt für Migration und Flüchtlinge 2011d) and housing. (Bundesamt für Migration und Flüchtlinge 2008h), education (Bundesamt für Migration und Flüchtlinge 2008e), language ac-

## Towards a Hegemonic Understanding of Integration

Not unlike the Migration Reports discussed in the previous chapter, initial research projects in the field of integration research were characterized by a strategy of mapping out the "unknown terrain" of integration. As one BAMF researcher noted, around the time of the foundation of the Research Group, integration was considered *terra incognita* for the Federal Office:

> "If I look back on the time when I started working here, integration was the great new thing. It came into being with the new immigration law and was defined as a task for government for the first time."[89]

While this statement represents a quite common point of view in the state administration and academia, it has to be historically contextualized. Given the long-standing tradition of municipal integration policies on a local level, this statement seems to be a slight exaggeration. Many of the later discussed and implemented tools of integration policy, such as language courses, institutional support networks, or educational programs have been developed and implemented since the late 1970s. In operative terms, integration policy-making in the 2000s did not differ extremely from similar policies in earlier times. However, there are two core differences in regard to the political legitimization and the theoretical foundation of integration policy. The first difference can be connected to the "paradigm shift" following the abolishment of the "no country of immigration dogma" as described in Chapter 2. Secondly, from a scientific perspective, the overarching principle of "no immigration" hindered the development of a coherent understanding of what integration actually was to achieve.

In the "no country of immigration" paradigm, integration was (somewhat paradoxically) understood as a temporary relief of social hardship before the

---

quisition (Bundesamt für Migration und Flüchtlinge 2008f), media use (Bundesamt für Migration und Flüchtlinge 2010a) and political participation (Bundesamt für Migration und Flüchtlinge 2012g). The Integration Panel is a longitudinal study among integration course participants whose results have been published in various working papers and research reports between 2009 and 2013. See Bundesamt für Migration und Flüchtlinge 2011a for a full report of the results.

89 "Wenn ich mich zurück erinnere an die Zeit, als ich hier angefangen habe, [...] da war eigentlich Integration der große, neue Begriff. Der mit dem Zuwanderungsgesetz gekommen ist und zum ersten Mal wirklich auch im Sinne von einer staatlichen Aufgabe definiert wurde."(Interview with a BAMF researcher, 2015)

eventual return of the foreigner.[90] Therefore, there existed no need for a generalized concept of integration as a social process, there was no reason to analyze its components, its aims, its effects, and how to describe it in an abstract way. In these respects, the concept of integration was, in fact, "a great new thing" which the BAMF researcher was referring to; given the legal changes of the Residence Act in 2005, it became necessary to define both the theoretical understanding of integration as well as the concrete policy measures connected to it.

Given the specific connection between the BAMF's central responsibility for integration policy-making, the Research Group saw itself in a prime position to contribute to the formulation of a theoretical concept of integration. Not only the BAMF, but almost every other involved state actor such as municipalities, the Federal Government, and the Commissioner for Foreigners worked on a definition of what integration actually meant. Between 2003 and 2010, every larger municipality issued an "integration concept,"[91] the Federal Government worked out an "Integration Plan" in 2007, and a working group of representatives of the Federal States put forward similar efforts.[92] Between these concepts, a quasi-hegemonic theoretical understanding of integration was formed which consists of three main elements: first, theoretical references to Hartmut Esser's assimilation theory; second, the ideal of equal participation of both migrants and autochthon population; and third, instruments of political steering and indicator-based monitoring from the "New Public Management" toolbox.

On a municipal level, the 2003 competition titled "Integration is No Coincidence", organized by the Ministry for the Interior together with the Bertelsmann Foundation, promoted the formulation of municipal integration plans. The basic idea behind this competition was a focus on concrete policy tools and their implementation. Most importantly, successful integration was to be achieved with the introduction of New Public Management governance techniques, such as best-practice models and the principle of indicator-based policy steering. Overall, the strategy of framing integration policy in a decidedly technocratic, somewhat apolitical language is clearly discernible.[93] Most of the integration concepts contain a basic definition of the aims and purposes

90 Lanz 2007, p. 252

91 Friedrich and Waibel 2012, 67f.

92 Cp. Worbs 2010

93 Cp. Friedrich and Waibel 2012, p. 67

of integration policies. Most commonly used definitions refer to an understanding of equality, understood as "equal participation of migrants in society, economy and culture"[94] which is usually achieved in a two-sided process, including responsibilities for both migrants and non-migrants.[95] The common denominator of most of these definitions is the desire to formulate an abstract, general, and politically uncontestable understanding of integration. The working definition put forward by an official of the Federal Statistical Office can be seen as paradigmatic of this understanding of integration and integration measurement in this respect:

> "[W]e discern migrant and non-migrant populations according to specific socio-demographic, socio-economic variables: education, labor participation, unemployment, income. This is how we measure integration. This is an absolutely unemotional account, not 'assimilation versus growing together.' Rather, we simply analyze if two parts of the population are different in respect to their socio-demographic features, or not."[96]

The theoretic background of this understanding of integration was developed by sociologist Hartmut Esser in the 1980s in his comparative studies of foreigners from different former "Guest Worker" countries, most importantly Turkey and Yugoslavia. Esser conceptualized integration as a process of social learning, consisting of a series of individual investment decisions into mainly cultural and social capital. For example, a German school diploma or German language skills are conceptualized as investments into "German" cultural capital, whereas marriages with a spouse of the same ethnic background, a non-German diploma or an apartment in an ethnic neighborhood are conceptualized as investments in "foreign" social capital. Esser differentiates between

94 Beauftragte der Bundesregierung für Migration, Flüchtlinge und Integration 2005, p. 182

95 Unabhängige Kommission "Zuwanderung" 2001, p. 204

96 „[W]ir unterscheiden die Migrantenbevölkerung und die Nicht-Migrantenbevölkerung bezüglich bestimmter soziodemographischer, sozioökonomischer Variablen [...], Bildungsabschluss, Erwerbsbeteiligung, Arbeitslosigkeit, erzieltes Einkommen. So messen wir Integration. [...] Das ist eine völlig unemotionale Darstellung, [...] nicht Assimilation versus Zusammenwachsen. Sondern, es wird einfach nur analysiert, ob sich die zwei Bevölkerungsteile bezüglich der Soziodemographie unterscheiden oder aber nicht."(Interview with a researcher at the Federal Statistical Office, 2016)

four dimensions of integration: structural, cultural, social, and emotional.[97] These four dimensions are often used as indicator categories for empirical research on integration, municipal indicator sets for integration monitoring, or other governmental integration policy documents. This is even true for concepts which do not directly refer to Esser's theory but rather use the category order of Esser's integration dimensions without a reference to the source.[98] Out of the four categories, the cultural dimension stands out since it is often prominently discussed. Coincidence or not, this dimension refers most directly to individual learning efforts: these include above all the acquisition of language, but also more generally the acquisition of context-specific social and normative knowledge.[99] Structural integration is measured by the participation rate in various relevant social systems, such as the education system as well as the labor and housing market. In the dimension of social integration, the degree of social interaction between the indigenous and the migrant populations is under scrutiny, often operationalized by measuring the share of bicultural marriages, or the share of Germans among friends and acquaintances, or the practice of giving "German" versus "foreign" names to children. The last dimension, emotional integration, describes the "development of a certain emotional loyalty [...] and the association with one's own identity."[100] While this dimension is usually the least well-developed in regard to indicators, often the naturalization quota is used in this context.[101] The borders between these dimensions are somewhat in flux; indicators are sometimes regrouped from one category to the other.[102] Furthermore, there is no clear hierarchy between integration dimensions. To be sure, there seems to be a consensus that emotional identification represents the successful conclusion of integration, whereas cultural integration, most importantly language acquisition, is seen as the starting point of the process. However, most scientists agree that all dimensions of integration precondition each other to some degree.

97 Esser 2001, p. 73

98 Friedrich and Waibel 2012, p. 57

99 Some integration indicators use the total fertility rate of women in this context. Cp. For example Magistrat der Landeshauptstadt Wiesbaden 2012, p. 20

100 Esser 2009, p. 87

101 Magistrat der Landeshauptstadt Wiesbaden 2012, p. 20

102 Interview with a BAMF researcher, 2015

In summary, after 2005 there have been considerable efforts to shape the new field of integration policy-making from a theoretical-abstract perspective by different state actors. The emerging understanding of integration as a structural similarity between migrants and non-migrants can be considered hegemonic and somewhat naturalized. BAMF-publications meanwhile refrain from direct references to Esser's theories and present the integration dimensions as self-explanatory sub-chapter headlines in their research reports.[103] For example, one BAMF researcher presented the application of Esser's theory as a somewhat natural choice without alternatives:

> "At the time, there were almost no other concepts. Esser is one of the first who has conducted systematic integration research and has tried to conceptualize it, and to transfer it to the German context. In a way, it is alternativeless. These four dimensions of integration, as a specific, pre-defined structure, constitute a well-probed principle."[104]

Another factor which comes into force specifically for the Research Group staff is the fact that some of the BAMF researchers, especially in the earlier years of the Research Group, obtained their academic training under Esser's professorship at the university in Mannheim.[105]

## Practical Relevance: From Migrant Assimilation to Migration Management

Up to this point, the general development of integration research after its establishment as a "guiding principle"[106] has been outlined: the hitherto "un-

103 Cp. also Eichenhofer 2013, 195 ff.: by discussing the intellectual history of integration theories, Eichenhofer refers shortly to selected "classic" (Durkheim, Weber) as well as "modern" (Habermas, Parson) theories, which he then discards as "undercomplex," "not implementable," etc. Some elements of Esser's discursive apparatus (such as the four integration dimensions, social- and system integration, etc.) are then introduced as "classic", without directly referring to Esser.

104 "Es gab damals auch keine anderen. Esser ist [...] einer der ersten, der wirklich systematisch Integrationsforschung betrieben hat und versucht hat, das zu konzeptionalisieren, oder auch auf den deutschen Raum zu übertragen. Es ist sozusagen alternativlos. [...] Diese vier Dimensionen der Integration, [als] eine gewisse Struktur, die damit einfach vorgegeben worden ist [sind] ein Prinzip, das sich bewährt hat."(Interview with a BAMF researcher, 2015)

105 Interview with a BAMF researcher, 2015

106 Gusy and Müller 2012

known terrain" has been mapped out with the help of Esser's assimilation theory and New Public Management[107] policies to overcome the decade-long deadlock in this policy area. In the next paragraphs, the discussion concentrates on the BAMF's research output and specifically its political usefulness. In this, a distinct shift is discernible from an initial strategy of "mapping out" to the consolidation of a specific, assimilationist understanding of integration.

As demonstrated above, the original research impetus consisted of the idea to construct an integration monitoring system similar to migration reporting.[108] In this context, one of the most important long-term research projects in the BAMF's integration research was production of the "integration reports." The project was started in 2006 with the aim of an "all-encompassing, data-based presentation of the status of integration of immigrants to a wider public."[109] Initially, the integration reports were conceptualized in a very similar fashion as the Migration Reports. Reporting was to consist mainly of compiling statistics from different governmental and academic resources and then presented in biennial reports. In this context, integration policy-making in this understanding is a prime example of a knowledge-based policy style which implements political measures in accordance to a scientific concept of integration. The scientific understanding (of successful integration) is operationalized with indicators which promise to deliver technically well-made policies untainted by ideology. In this context, the question of the practical use of the knowledge created at the BAMF is paradigmatically instrumental: knowledge is needed for defining criteria for successful integration, and for associated indicators; policy-making is a somewhat mechanistic effect of this scientific knowledge since it is thoroughly governed by enhancing or lowering certain indicators; policy evaluation is likewise simply achieved by monitoring the according numbers.

However, integration research took a different direction as of 2008 as is visible in a change in the publication strategy of the integration reports: instead of compiling data in successive reports, single and relatively long Working Papers were published on commonly discussed integration-related topics such as language acquisition, ethnic segregation in housing, schooling,

107 For an overview over New Public Management principles in integration and migration policy making, see Geiger 2013, p. 25

108 Cp. for example Worbs 2010

109 Bundesamt für Migration und Flüchtlinge 2009b, p. 31

and labor market aspects. Furthermore, there seems to be a distinct research strategy in the reports to fill the data gaps in integration reporting: Especially in its earlier publications, the Working Group has published integration reports on "soft dimensions of integration"[110] which are commonly underrepresented in research: these include inter-ethnic contacts, media use, and political participation. In some of these studies, uncommon data resources are analyzed, especially in the earlier volumes of the integration report series (for example, housing and car ownership data). This could be read as a distinct strategy to broaden the data basis of integration research to represent the four integration dimensions more equally in indicators.

Despite these efforts, the concept of statistical monitoring similar to Migration Reports was subject to a steady decline: after the initial change of publication strategy in 2008, the integration reports were terminated in 2012. The reasons for this change in strategy are not specified; in the literature, political resistance from institutional rivals is mentioned in this context: according to Boswell, the commissioner for foreigners considered the integration reports a transgression of its own integration report series and thus mobilized some political resistance.[111] Additionally, the decline of research efforts in this direction could be connected to inherent contradictions of the concept of integration monitoring as a knowledge-driven, technocratic and apolitical management system. This understanding of policy-making became problematic both in regard to the theoretical groundwork and the actual implementation: concerning the latter, initial optimism of connecting policy-making with scientific research gave way to the insight that the connection between indicators and policy objectives is in many cases less straightforward than initially expected.[112] Concerning the theoretical groundwork, the abstraction of a coherent, politically uncontested understanding of integration was hard to achieve. Most importantly, two interconnected reasons are named in the literature: a lack of data, and a lack of common political interests. The latter point was connected to a somewhat deceptive hope that a common understanding of integration was within reach once the ideological dispute over its aims was set aside. However, even if this de-politicizing could be achieved, a diversity of interests still pertains: a communal integration concept somewhat naturally emphasizes questions of ethnic segregation of neighborhoods whereas

---

110 Bundesamt für Migration und Flüchtlinge 2008b, p. 42

111 Boswell 2015, p. 25

112 Directorate-General Justice, Freedom and Security 2010 quoted after Worbs 2010, p. 6

this factor does not play such a strong role for larger political units such as the federal state. Partly connected to this is the varying availability of data; most complete data sets can be found mostly in structural areas such as labor market and education, whereas data on identification, cultural aspects, and other important fields are rather patchy. Moreover, both problems worsen according to the level of administration: while most municipal actors are able to define integration policy aims and can link them to data, this process is more problematic on a federal level, and in extension, for the BAMF. Most federal statistics are produced according to a citizenship-logic, therefore rendering naturalized migrants invisible to integration reporting. In consequence, not only the BAMF, but also the Federal Commissioner for Migration, Integration and Refugees has meanwhile abandoned the plan to establish an indicator-based measurement system for integration policy.[113]

However, despite the failure of state-level integration monitoring, it can be argued that the initial impetus to create a technocratic, non-ideological policy field in the area of integration succeeded in different ways. This success can be described by what Ferguson called "depoliticization": by framing political problems in technical terms, and providing according solutions "to which no one can object," originally political questions are removed from the political realm to be administered by the state bureaucracy.[114] This depoliticization has a twofold effect. First, as already described, it transforms the object of intervention into a merely technical question. This includes not only the alleged targets of integration measures, but also the general political environment. As one interlocutor noted, the governmentality of integration contributed to a more positive recognition of the Federal Office's work, especially among support NGOs who used to largely be in principal among the opposition:

> "the better image of the BAMF [...] has been massively influenced by the fact that since 2005, the BAMF has distributed millions of Euros for integration projects to non-government organizations via project funding. [...] this is a fundamental difference, to conduct integration policy and to support it financially, to establish public relations, and to cooperate with civil society organizations."[115]

---

113 Sachverständigenrat deutscher Stiftungen für Integration und Migration 2017

114 Ferguson 1994, p. 256

115 "das verbesserte Bild des BAMF ist [...] ganz massiv darauf zurückzuführen, dass seit 2005 das BAMF die Behörde ist, die Millionen [...] von Integrationsfördergeldern verteilt hat an Nichtregierungsorganisationen über die Projektförderung [...] Das ist

Second, depoliticization supports the expansion of bureaucratic organizations which are carrying out technical solutions to the identified problem. This can again be illustrated by the mechanism of distributing funds for integration projects, most importantly integration courses: upon receipt of project funds, NGOs have to comply with the BAMF's rules of project funding and documentation. Sanctions can be carried out not by way of punishment, but by way of non-renewal of short-term project cycles. In both respects, the BAMF has been quite successful in introducing a technical understanding of integration, implemented as a bureaucratically steered process in which migrants are compelled to display individual effort in integration policies.

Coming back to the question of political usefulness of integration research, the shift of knowledge production from an indicator monitoring system becomes plausible in this context: once the policies are established, foundational research, the establishment of an overview or the completion of data for the establishment of an indicator system is not needed anymore. On the contrary, report series rely on a stable framework of data, indicators, and theory, so this built-in tendency of inertia favors the continuation of once-established patterns which do not need to be reconfirmed (or worse: questioned) by research. Instead of the completion of an "all-encompassing overview," as implied by the Integration Report series, the logic of practical applicability required different research projects after the basic principles of integration policy-making and the related theoretic concept have been established. Instead of foundational knowledge, minute, technical knowledge about specific integration policies is needed. This shift is visible in the integration panel project as mentioned above, which focuses on the success of integration course participants, therefore conceptualizing integration not as a societal process but rather an outcome of individual effort in the context of government policy. This shift is in line with according demand for technical knowledge on the side of the administration, as Boswell noted: especially the BAMF integration unit increasingly commissioned studies on technical issues such as, for example, the effects of particular teaching methods and incentive systems on the success and participation rates in integration courses.[116]

---

schon ein fundamentaler Wandel […] im Auftrag des BAMF eben auch Integration zu machen, zu fördern, nach außen zu gehen, sich zivilgesellschaftlich zu verankern." (Interview with a former BAMF researcher, 2016)

116 Boswell 2009b, p. 174

This shift in demand for knowledge seems to be the decisive factor in the adjusted output of integration research in the BAMF from about 2009/2010 on. Practically relevant knowledge in the context of integration was therefore subject to a shift from a rather generalized perspective, not unlike the Migration Reports, to a more technical, specialized research design. "Practically relevant" meant, at the beginning of the research, activities to contribute to an abstract understanding of integration, which was mainly interpreted through Esser's assimilation theory, a concept which has gained widespread application in governmental contexts. Over the years, this concept was, however, specifically reinterpreted to suit the context of depoliticized bureaucratic administering of integration according to which integration was conceptualized mainly as an outcome of migrant's individual efforts.

Over the years, the hegemonic understanding of integration became gradually more differentiated. After about 2012, a second stream of integration studies, connected primarily with different forms of temporary labor migration, emerged.[117] Two features of the integration concept of these labor-market related studies are of particular interest. First, they lack the unified theoretical framework of reference as in the case of Esser-inspired integration studies. Second, the target groups in question here – highly-qualified immigrants, self-employed migrants, and university graduates – have a rather privileged status in common and are all perceived as economically useful. All in all, the studies are less rigidly structured and seem less directly oriented towards individual negative features of migrants which have to be corrected by policy intervention. Integration in these cases is less an outcome of individual effort and integration policy measures. Rather, it is conceptualized as the outcome of a combination of personal features and structural conditions of the labor market.[118] This interpretation also explains the selective application of

117 Cp. Bundesamt für Migration und Flüchtlinge 2009e, Bundesamt für Migration und Flüchtlinge 2012i, Bundesamt für Migration und Flüchtlinge 2014a

118 This difference becomes even more pronounced when analyzing labor-market related studies which do not focus on privileged target groups, for example Bundesamt für Migration und Flüchtlinge 2011d or Bundesamt für Migration und Flüchtlinge 2011e: In this context, labor-market aspects of integration seem to be discussed within the individualistic approach: In these studies, migrants represent rather a threat to social order than an opportunity for economic development; Migrant economic activities are discussed in predominantly negative contexts such as unemployment, low wages, or unsafe and unstable employment conditions. Structural factors for these problems are however rarely discussed (apart from one mentioning of discriminating migrant

Esser's concept to specific target groups such as immigrated spouses, Muslims, and integration course participants: all of these groups are attributed one deficit or the other, mainly a lack of integration, but also (especially in the case of Muslims) problematic ethical values.[119] The integration of these target groups is perceived as a correction of specific deficits; integration policy in this context usually comes in the form of educative measures.

To summarize, governmental integration research was initially characterized by the idea of introducing a technocratic policy-style of integration steering as an answer to decades of political backlog. In practice, however, this ideal of a knowledge-based policy has never been implemented; the end of the according project of an "integration report" monitoring integration success can be considered a consequence of this fact. In short, the notion of integration was gradually reformulated from a project for the whole society to a government-organized education of migrants. What started out as a "two-sided process" which entailed "responsibilities for both migrant and autochthon population" was translated into a set of policy-tools which target exclusively migrants.[120] This means that immigrants were required to put their individual efforts into integration measures, whereas responsibilities for the indigenous population were transferred to the state which designs and offers integration measures. The main policy tool in this context, the integration courses, mirrors this understanding very well: these courses were designed essentially as a language course with a short part on civic-historical education. These courses are an educational measure designed to counter migrant deficits – be it lack of information, lack of language skills, or lack of socio-political or historical knowledge – which are believed to hinder migrants' full participation in the German society. This deficit perspective and the cure to it is a well-pronounced feature of foreigners' education which developed during the 1970s and 1980s as described in Chapter 2. Despite the claim of a "paradigm change," the actual integration course design and contents bear many similarities to earlier "Guest Worker" language courses as well.[121] In-

youth art the vocational training market, see Bundesamt für Migration und Flüchtlinge 2011d, p. 33).

119 For example, a comparative study on gender roles of Muslim and Christian immigrants explicitly names the identification of a "need of support of the equality of genders" among Muslims Cp. Bundesamt für Migration und Flüchtlinge 2014c, p. 5

120 Cp. Wimmer 2009, p. 332

121 Cp. Zur Nieden 2009

tegration research nicely demonstrates how knowledge and political usefulness considerations influence each other and change over time: while integration indicator systems were clearly developed out of a perceived need for an overview of an unknown phenomenon, later research was to abandon this all-encompassing perspective and focus on technical, non-political aspects of integration policy-making, thus creating an understanding of integration as an outcome of the migrant's individual effort. This hegemony was reduced with the emergence of newly created immigration channels for privileged migrants which required altered, less assimilation-oriented integration knowledge. In the next chapter, the discursive effects of this segmented knowledge structure will be discussed.

## Effects on the Knowledge: Selective Blindness towards Discrimination

In the last paragraphs the selective application of Esser's theory to research projects has been documented: the main finding is that Esser's theoretical understanding of integration is reinterpreted from a general theory of integration to the monitoring of migrants which are subject to governmental intervention by way of integration policies. In this chapter, the process of selectively applying theory to governmental knowledge will be scrutinized in some detail. The point of departure is the above-mentioned observation that Esser's concept is not uniformly adopted in BAMF research; there is a well-visible shift towards individual factors of integration. The main hypothesis will be developed according to which this selective application is responsible for a bias in governmental integration research which renders knowledge selectively blind towards structural discrimination of legally less privileged status groups.

The selective application of Esser's theory is a known structural feature of the BAMF's research; however, researchers do not problematize the fact. In regard to the well-pronounced stress on structural factors of integration monitoring, BAMF researchers justify this selective interpretation with the pragmatic style of knowledge production: the argument goes that not theoretical coherence, but practical applicability is the main yardstick for quality of the research.

> "We conduct commissioned research [...] for the Ministry of the Interior. The research questions stem less from the academic world but rather here from the Federal Office and the Ministry of the Interior. For this commissioned

research, for answering the questions, Esser's concept suits itself very well. This is why we have used it, but we are not in any way committed to it."[122]

This emphasis is in line with other integration-related bodies of knowledge as discussed above, for example municipal integration reports: these, too, put a focus on the structural dimension, whereas the emotional/identificatorial dimension is somewhat under-represented. Often, the lack of adequate data is presented as the decisive reason for this.

However, while this might be true, the degree of disregard for structural barriers to integration (i.e., discrimination and racism) is extreme in the BAMF's case: between 2005 and 2015, *not one single* BAMF research project deals with this topic as a major focus.[123] This neglect becomes even more apparent when BAMF research is compared with academic knowledge production, where studies on discrimination/racism feature among the most important research topics in migration studies. If only lacking data were the problem, the difference between academic and governmental knowledge production were not as large, since both would have to struggle with the same limitations. To be precise, this does not mean that the BAMF completely disregards the topic; however, it only appears as a sub-topic in some publications. In the next paragraphs, the specific governmental reading of discrimination will be discussed using the Integration Panel, one of the most prestigious long-term projects of the BAMF.[124]

In the Integration Panel, the progress of language acquisition of integration course participants is scrutinized against a control group controlling several individual socio-economic indicators such as age, gender, education, la-

---

122 "Wir machen Auftragsforschung [...] für das Innenministerium [...]. Die Fragestellungen kommen ja weniger aus dem universitären Bereich, als hier aus dem Amt, plus dem Innenministerium. Für diese Auftragsforschung, für die Beantwortung der Fragen, [...] eignet sich das Esser'sche Konzept sehr gut. Darum haben wir es verwendet, aber wir sind nicht irgendwie [daran] gebunden."(Interview with a BAMF researcher, 2015)

123 As of 2015, no research project was explicitly dedicated to studying discrimination; accordingly, only 5 out of more than 500 external publications (counting lectures, workshops, and book publications) deal explicitly with the topic of discrimination. (Own evaluation of data from yearly research reports).

124 The Integration Panel was a longitudinal study on the integration success of integration course participants, spanning over publications from 2008-2013. Cp. Bundesamt für Migration und Flüchtlinge 2011a

bor market participation and also discrimination.[125] Concerning the latter, a table shows that between one-quarter (integration course participants) and one-third (control group) of the respondents "feel discriminated against" either "sometimes" or "often." These numbers are somewhat optimistically presented as good news: "a majority of the respondents does not feel discriminated because of their ethnic belonging." However, two sub-groups of the respondents report discrimination much more often than the average: first, members of the control group, and second, members of visible minorities, operationalized as "migrants form sub-Saharan Africa [...] and Asia," from the study group. With 29% of the respondents of the first and 49% of the latter group, the respective numbers are significantly higher than average. In the discussion of the data, there seems to be an honest effort to analyze the connection between integration and discrimination, since "an open attitude of the receiving society is important for emotional integration" which can be hampered by "perceived discrimination." However, the analysis of the data is not geared towards this connection between integration success and discrimination. Instead, it remains firmly grounded in the individualistic approach: by analyzing the differences between former course participants and the control group, the report states that the differences in perceived discrimination do not necessarily have to be caused by factually higher levels of discrimination, but rather just by according perceptions: Simply put, course participants could be less likely to "feel" discriminated. The argument goes that "Perceived discrimination depends on the subjective feeling, which means, the specific disposition to interpret social conflicts as ethnically or culturally motivated." This is bolstered with data from a study among Turkish and Yugoslavian youth: by quoting numbers which suggest that respondents expect discrimination more often than actually fall victim to it, the study argues that a good share of the problem lies in the overly pessimistic attitude of some migrants. The underlying argument originates from the individualistic approach: discrimination is consistently referred to as "perceived discrimination," framing it somewhat as a private matter instead of a structural barrier to integration. In the same context, the report mentions a negative correlation between discrimination and social contact with autochthon Germans. Again the language is rather tentative and cautious, formulating an optimistic expectation that reports discrimination diminishes once friendly contact with Germans cancels out eventual discrimination experiences. In a similar way,

125 All quotes in this paragraph Bundesamt für Migration und Flüchtlinge 2013b, p. 74

lower levels of discrimination experiences among integration course participants are tentatively interpreted as an outcome of the "positive recognition" through the integration course. However, despite the positive language, discrimination is portrayed as the result of poor understanding of migrants who wrongfully interpret usual conflicts as ethnically motivated and thus "feel" discriminated. All in all, it seems that discrimination is most of all a problem of the non-integrated migrant.[126] Thus, the cause and effect relationship between discrimination and integration is in a way reversed.

In conclusion, there seems to be a structural contradiction at work when discussing discrimination in the framework of the Integration Panel's analysis. On the one hand, there is an honest effort to include a perspective of structural barriers to integration into the analysis; again and again, the importance of an "open society" is stressed. On the other hand, however, discrimination is not conceptualized as a societal barrier to the success of language acquisition. This is above all caused by the individualistic framework of analysis which supports an image of integration as the result of the individual effort on the side of the migrant. This design clearly reaches its analytic confines: The plain hypothesis that visible minorities are much more likely to be singled out for discriminatory acts is impossible to capture with the individualistic framework. The same is true for the reference to discrimination as a "feeling" or "perception." This conceptualization renders discrimination an individual feature of the migrant instead of a societal problem with structural impact on integration. This perception is argumentatively connected to a lack of integration of the migrant. Compared to the introductory statement about the importance of an open society for integration, the cause-effect relationship is reversed: in this sense, discrimination is not a factor which hinders integration, but rather, a lack of integration effectuates higher levels of "perceived discrimination" especially by the (incorrect) interpretation of social conflicts as ethnically motivated. The result seems to suggest that discrimination can best be cured by enhanced integration efforts.

The above-mentioned effect of selective application of different frameworks of analysis according to the logic of perceived usefulness of different legal groups is most clearly visible in the respective conceptualization of discrimination. The individualistic approach based on Esser and the consequen-

---

126 Cp. also Bundesamt für Migration und Flüchtlinge 2013c, 134f.: In this study," discrimination experiences" is mentioned as an indicator for the "perceived acceptance of migrants on the side of the majority."

tial downplaying of structural factors can be set off by what has been called "Welcome Culture" in the context of the integration of highly-skilled migrants. The term was introduced around 2005 by employer associations which essentially called for a preferential treatment of skilled migrant workers by the state bureaucracy and the society.[127] By 2012, the BAMF had assembled a "Round Table Welcome Culture" and published recommendations for the conduct of the state administration vis-a-vis immigrants, for the inclusion of immigrant organizations, and for other "best practice" examples. Notably, a direct connection is drawn between discrimination as a relevant factor for society on the one hand and integration difficulties on the other:

> "According to a study [...] on welcome culture in Germany, around 65 percent of the respondents connect additional problems and conflicts with immigration. These opinions among the German receiving society can result in experiences of discrimination among migrants, which influence the subjective or actual readiness for integration negatively. All in all, there is a lack of welcome culture and positive recognition of diversity in Germany."[128]

In the BAMF's research work, this concept of "positive recognition" is visible as well. Notably, a researcher mentioned such a notion in the context of migrants' language acquisition: Esser argues that learning German is essential for integration since command over German is a prerequisite to entering the labor market. While this position is relatively common-sensical, Esser's negative stance towards the migrant's language of origin is not: based on the conceptualization of language acquisition as an investment in either "German" or "foreign" cultural capital, Esser constructs a zero-sum-game that any investment in "foreign" social capital automatically entails less investment in "German" capital, which is considered harmful for integration. This, however, is refuted by linguists which present contradicting data.[129] The problem is that Esser's approach is not empirical, but rather theoretical on the basis of economic utility: most languages of origin are not important to the labor market and thus he argues that the formal training in the language of origin is in the best case not harmful but can never be considered an advantage for the acquisition of German. Given this academic controversy, it seems that the BAMF's

---

127 For an overview over the different iuses of the term, see Haller 2017, 137f.

128 Bundesamt für Migration und Flüchtlinge 2013d, p. 3

129 Hetfleisch 2013, 226f.

position towards this particular element of Esser's theory is changing in the light of the "welcome culture" debate:

> "So [Esser] concludes that immigrants have to learn German in any case, which is a precondition to enhancing job chances which is part of successful integration. In this context, command over multiple country of origin's languages is irrelevant and therefore not so important. However, our position is that this cannot be put so simply, that the human capital is important no matter what, for the person and for integration as well."[130]

This is to a degree mirrored in the BAMF's research on integration. Since about 2012, the formerly hegemonic position of an Esser-inspired framework of analysis as described above has changed, since it is no longer applied uniformly to all integration research projects. An increasing focus on structural factors of discrimination is discernible in the BAMF's labor-market related studies. In an early study on highly qualified immigrants in 2009, the matter is vaguely defined as "general life satisfaction."[131] In the 2014 study on university graduates, the topic is discussed more extensively, following broadly the main areas of concern of the "welcome culture" white paper: satisfaction with the state and university administration, access to information, and the general "feeling of being welcome."[132] Notably, the fact that more than 40% of the respondents criticize the short duration of their status title is prominently discussed in the study; this in turn reflects the fact that legal regulations are included in the analysis of integration, a notable difference to Esser-inspired studies.[133] This being said, it does not mean that Esser's theory has been replaced by a more open and discrimination-sensitive framework, but rather, by a selective logic of economic usefulness. The analysis of structural factors

130 "darum schlussfolgert [Esser], die Leute, die kommen, müssen auf jeden Fall Deutsch lernen, weil nur dann sind die Erfolgschancen auf dem Arbeitsmarkt überhaupt gegeben und zu einer erfolgreichen Integration gehört das eben dazu. Und die [...] Beherrschung von mehreren Herkunftslandsprachen ist dafür irrelevant und deshalb ist es nicht wichtig. Und, da haben wir [...] die Position, dass das nicht so einfach gesehen werden darf, dass dieses [herkunftslandbezogene] Humankapital, [...] trotzdem für die Person wichtig ist, und auch für die Integration."(Interview with a BAMF researcher, 2015)

131 Bundesamt für Migration und Flüchtlinge 2009e, 73ff.

132 Bundesamt für Migration und Flüchtlinge 2014a, 230ff.

133 Ibid., p. 10

of discrimination is conducted only in the context of economically useful migrants such as highly skilled workers and university graduates.[134]

To summarize, the shifting and selective conceptualization of discrimination highlights the structural confines of governmental knowledge production as present in the BAMF quite well: practical relevance requirements predispose a shift of research activities from fundamental research and theory development to the research of integration processes especially of those target groups which are the object of political interventions. In this context, some built-in features of the BAMF's representation of Esser's integration theory as well as the deficit-orientation of integration policy-making can be made responsible for a well-pronounced stress of individual factors and above all a disregard for discrimination as a structural barrier to integration. In the example of privileged immigrant groups, the notion of integration is different: individual and structural factors are analytically more balanced, so that successful integration is conceptualized as a result of both structural opportunities and individual features of the migrants. This can be seen to disprove the assumption that the BAMF fully disregards the topic, as both the quantitative analysis of research project topics and the data from SOLIS as discussed above might wrongfully suggest. Rather, as the discussion of "welcome culture" and Research Group publications on privileged migrant groups show, it seems that the BAMF successively takes up criticism and incorporates it into its knowledge production. This criticism, however, is subject to the same practical applicability considerations as the rest of the body of knowledge.[135] Therefore, critical knowledge is allowed only in a "productive sense" if it can be used to improve the overall order of knowledge without targeting its principles such as the methodological framework of Esser's integration theory. In this sense, "welcome culture" is interpreted as a concept for supporting the integration of immigrants who are regarded economically useful; the (historically older) understanding of integration as an individual effort in the framework of government policy remains in place for all other immigrant groups. This cognitive order is a direct outcome of the requirements of practical applicability, where in the area of temporary work migration, the policy aims are different from other policy areas such as resettlement, asylum, and family reunification: the governmental logic is recruitment of qualified workers

134 Bundesamt für Migration und Flüchtlinge 2013d, p. 6

135 Mecheril et al. 2013, 30f.

in the former case and education of migrants in the latter. This entails a difference in the targets of governmental actions: intervention policies naturally target individuals, while support includes also structural barriers such as legal requirements to status titles or experiences of discrimination.

## Conclusion

In this chapter, the development of a theoretical understanding of integration has been analyzed. Mainly, three phases of development can be discerned: first, the concept of an indicator-based monitoring system for integration until about 2008; second, the hegemonic application of Esser's assimilation theory; and third, the gradual diversification of integration concepts according to the economic usefulness of the to-be-integrated subject since about 2012. Regarding the first phase, research interest was based on the one hand on the search for an abstract understanding of integration; on the other, the goal was to make integration an orderly and legible field of policy-making through statistical reporting. While the technocratic elements of steering were largely abandoned, a widespread, if not hegemonic understanding of integration was successfully installed. This understanding, based on a modified version of Esser's assimilation theory, can be regarded as the intellectual basis for integration research. According to Hetfleisch (2013), the BAMF's understanding of integration is characterized by a principle of "methodological individualism" borrowed from Esser's assimilation theory. As has been demonstrated, this principle focuses on individual actions and systematically underrepresents systemic discrimination and racism; furthermore, the model of rational utility maximization equalizes successful integration with the economic utility of a given migrant. In effect, Esser's theory legitimizes a meritocratic world-view in which exclusion and discrimination are conceptualized as the outcomes of poor economic integration based on the lack of individual will or ability to assimilate.[136] Amir-Moazami's critique of the BAMF's study on Muslims can be regarded as exemplary for knowledge production which describes the BAMF's efforts quite well:

> "Although many of these studies make a studious effort to recur to the political rhetoric of 'bilateral integration', the main burden lies at the side of the respondents. [...] While integration is de-politicized, scientific research

136 Hetfleisch 2013, p. 227

> delivers the proof that integration is at the end the responsibility of the individual."[137]

While this understanding of integration can be regarded hegemonic not only for the BAMF, but also for most other state actors in Germany, there is a gradual reorientation and diversification in recent integration research. In the context of the rise of the "welcome culture" debate since about 2012, studies on single target groups are published which do not apply the one-sided, individualistic framework of analysis. In regard to the gradual diversification of research concepts, it is interesting to note how the understanding of integration changes according to the perceived use of the knowledge: in the case of migrants who are perceived as deficient in some way, integration is portrayed as the result of individual efforts within Esser's framework of analysis. In this context, practical applicability means that integration is primarily understood and analyzed through the lens of educational integration policies. In the case of temporary work migrants, international students, university graduates and other rather privileged status groups, integration is framed by balancing structural and individual factors of analysis. The fact that integration in these cases is conceptualized as an effect of structural opportunities can be connected to the governmental rationality of fostering the immigration of these status groups.

137 Amir-Moazami 2018b, p. 95

## Calming Public Debate through Objective Knowledge

One of the most successful studies of the Research Group was published in 2009 under the title *Muslimisches Leben in Deutschland* (Muslim Life in Germany, MLD). The research report analyzes data from a large-scale, telephone-based survey on Muslims in regard to their religious practices, and aspects of their integration. Together with some more research projects on different aspects of Islam and Muslims, these studies can be considered a core focus of research at the BAMF.[138] The knowledge produced about Muslims shares many characteristics with general integration research as analyzed in the last chapter, most notably the Esser-inspired theoretical framework of analysis. However, in regard to the conditions of knowledge production and the intended use of these studies, three core differences are apparent. First, all of the above-mentioned studies have been commissioned by the *Deutsche Islam Konferenz* (German Islam Conference, DIK), a forum of state officials and representatives of Muslim organizations founded in 2006. The direct connection between the research reports and a commissioning institution can be considered a core difference to more generalized integration research, which is usually either self-commissioned or mandated by the BAMF integration department. Second, it stands to reason that these different conditions of knowledge production are connected to changed practical applicability deliberations as compared to generalized integration research. As will be shown later in some detail, in the context of Muslims, research is structurally influenced by the aim to maintain quasi-neutral outsider perspective on Islam independent of the established "insider" and "security" discourses. In connection to this, the third difference to generalized integration research is the political usefulness of the knowledge: In the context of Muslims, knowledge is often geared towards the aim to maintain societal peace by disproving what are perceived as irrational, fear-driven, subjective, discriminatory and racist statements about Muslims.

In the following paragraphs, the BAMF's knowledge production about Muslims will be analyzed, focusing on the study "Muslim Life in Germany" as a paradigmatic example. In a first step, the study's background, its core concepts, methodology, and topics are presented, which serve as a basis for

---

138 Bundesamt für Migration und Flüchtlinge 2012e and Bundesamt für Migration und Flüchtlinge 2014c. In 2016, a second edition of Muslim Life in Germany has been published.

the analysis of its political usefulness by entertaining a narrative of sober, objective facts as a remedy against racist public discourses. The chapter concludes with a discussion of the epistemic quality of the knowledge mainly in regard to forming a specific image of the "Muslim other."

## Muslim Life in Germany

The origin of the study "Muslim Life in Germany" is on the one hand connected to the German Islam Conference, as already mentioned; on the other hand, it is embedded in an EU-wide trend of socio-demographic studies on Muslims since the beginning of the 2000s.[139] Concerning the former, it is worth briefly reconstructing the foundation of the Islam Conference as a policy tool in integration policy-making. The Islam Conference was founded in 2006 at conservative Minister of the Interior Wolfgang Schäuble's initiative. Schäuble's aspiration can be seen as a direct consequence of the "paradigm change" in migration policy-making as discussed in Chapter 3.1. In post 9/11 public debate, Muslims were frequently identified as a problematic group in respect to their threat to public security, their ethical values, or their socio-economic integration.[140] At the same time, politicians promoted a pragmatic recognition of a Muslim minority in Germany in replacement of the outdated "no country of immigration" dogma. In its original design, the Islam Conference was conceptualized as a policy tool to address both issues: on the one hand, it responded to publicly discussed integration deficits of Muslims; on the other hand, in the long term, it was meant to constitute something like an official political representation of all Muslims in Germany.[141] In the course of its existence, however, the Islam Conference developed similar to integration policy-making in general: what started out as a dialogue between equal partners in 2006 was gradually reformulated to a somewhat asymmetric education of Muslims or, as Tezcan (2010) described it, "a project of enlightenment."[142] According to Engler (2014), the seating order of the Islam Conference mirrors this development well:

> "The seating arrangement in the opening session of the Islam Conference provides a long table where 15 representatives of the German state sit vis-a-

139 Amir-Moazami 2018a, Schepelern Johansen and Spielhaus 2018, p. 125

140 Cp. Spielhaus 2013

141 Engler 2014, p. 67

142 Tezcan 2011, 94f.

vis 15 representatives of the Islam. [...] While the position of the state seems firm and enclosed, secured by the central position of the minister, this central place is vacant on the side of the Islam representatives. The seating arrangements [...] reflect the program of the Islam Conference in two ways. First, the spatial separation of the two parties coincide with the foundational separation of the two speaker positions 'German state' and 'Islam in Germany'. Second, by staging the collision between governmental unity on the one hand and Islam diversity on the other, the dialogue is turned into an examination. The Islam representative's role is to explain and justify, while the state representatives pose questions and make a judgment."[143]

In the course of the Conference, a dichotomy is constructed between well-integrated, enlightened, secular Muslims on the one hand and radical, conservative and anti-democratic Islamists on the other.[144] Accordingly, the topical framing of the Islam Conference focused strongly on a context of immigration, related integration problems, and constructed connections to homeland security issues.[145]

As already mentioned, the BAMF's studies on Muslims are part of a wave in socio-economic studies on Muslims starting around 2000. This surge in interest replaced a decade-long neglect of religious aspects in the immigration question.[146] This is well illustrated considering the BAMF's research agenda: until MLD, religion in general was not a major topic in BAMF research.[147] In the years after 2009, religious aspects in integration research are almost always connected to Muslims, either as the sole focus or comparatively with Christians.[148]

"Muslim Life in Germany" was at the time of its genesis one of the largest research projects conducted at the BAMF. For this study, respondents were selected from telephone book entries based on the criterion if their surname was common for selected Muslim countries of origin. Subsequently, about

143 Engler 2014, 83ff.

144 Müller 2018, p. 208

145 Engler 2014, 79f., Cp. also Müller 2018, 189f.

146 Amir-Moazami 2018a, p. 9

147 With the exception of two working papers on Jewish immigrants (Bundesamt für Migration und Flüchtlinge 2005c and Bundesamt für Migration und Flüchtlinge 2007b). These working papers however conceptualize Jewish immigrants through their legal status as contingent refugees, not as a religious minority.

148 Cp. for example Bundesamt für Migration und Flüchtlinge 2014c

6,500 telephone-based interviews were conducted.[149] This is a remarkable diversion from the usual routine at the BAMF, where data is usually collected in a pragmatic way: statistical data stems either from third sources or from respondents drawn from the AZR data base. In contrast to this, after a study on Ethnic Germans, "Muslim Life in Germany" was only the second empiric study for which raw data was collected. This elaborate methodology testifies to the unusually large employment of resources in connection to "Muslim Life in Germany".

The empirical part of the study is divided into three main chapters: after a discussion of socio-economic and migration-related features of the Muslim population, aspects of Muslim religiosity and religious practice as well as integration are discussed. Clearly, the most important question is the measurement of the number of Muslims in Germany.[150] As the BAMF notes, the main purpose of the study is to "determine the number of Muslims and their religious composition as precisely as possible"; every other question touched upon in the study – social issues, integration, and so forth – are portrayed as a dependent factor of this overarching goal.[151] This aim is justified with the fact that available data on Muslims was rather patchy and incomplete: in government documents, the size of the Muslim population was estimated by combining the numbers of foreigners from various countries with a Muslim majority. This estimate rendered 2.8 to 3.6 million persons.[152] The BAMF lists some reasons why this estimation method is problematic, most importantly a selection bias through migration (since religious minorities are more likely to emigrate).[153]

One important methodological question in this context was the definition of Muslim. This was operationalized in two steps: first, as mentioned, persons with a typical name from 49 countries with a significant Muslim minority

149 For a description of the methodology, see Bundesamt für Migration und Flüchtlinge 2009c, 36ff.

150 In 2016, the BAMF released an update of the number of Muslims. However, the database for the population register has changed in the meantime, which is why the two numbers are not comparable to each other. In 2011, the census revealed that the population registers exaggerated the number of foreigners in Germany by almost 15%; the number of Muslims is therefore inexact as well. Cp. Bundesamt für Migration und Flüchtlinge 2016c, p. 5

151 Bundesamt für Migration und Flüchtlinge 2009c, p. 20

152 Spielhaus 2013, 6f.

153 Bundesamt für Migration und Flüchtlinge 2009c, p. 20

were selected from the telephone book. Second, respondents were asked if they regarded themselves as Muslims or not. In short, a respondent would fulfill both criteria of possessing a Migrant Background from selected countries of origin and self-identification as a Muslim. The result was that between 3.8 and 4.3 million persons, or about 5 percent of the total population, self-identified as Muslims.

One should add that in the literature, some sources of bias in this method are discussed critically: in principle, telephone directories deteriorate in quality since they progressively cover an increasingly smaller share of the population; also, not every Muslim possesses a surname which makes him or her identifiable via the telephone book (especially convertites or married spouses which adopted another surname). On a conceptual level, critique targets the fact that only foreign nationals were included in the estimates, which reflects a common misconception of equating Muslims with migrants. Another problematic outcome of the study is the construction of so-called Muslim countries: on the one hand, in average only about half of the persons with a background from these countries self-identified as Muslim; in some of the allegedly "Muslim" countries the share of Muslims among the total migrants from that country was less than 10 percent (such as Kasakhstan, for example). on the other hand, persons from non-Muslim countries (in the logic of MLD) such as French, Dutch or US Muslims were excluded from the outset, thereby focusing on the legally least privileged share of the Muslim population.[154] Nevertheless, the study enjoys relatively widespread acceptance among scientific and political audiences. Almost all governmental publications, Muslim organizations, and scientific studies recur to these numbers, testifying to its widespread use. In the academic literature, the method is usually presented as relatively sound, resource-intensive as well as without viable alternative.[155]

The research report dedicates one chapter to the analysis of Muslim religiosity. Here, by and large, two main topics can be identified. In the first half of the chapter, religiosity is mainly captured and analyzed through a self-assessment of the respondent's degree of religiosity and the execution of ritual religious acts; this includes the observance of religious rules such as fasting, obeying religious food restrictions, attending religious services, and praying. This portrait is completed with a collection of publicly discussed pathologies connected to Muslims. These include the non-attendance to public schools,

154 Cp. Hernández Aguilar 2018, p. 34

155 Spielhaus 2013, p. 12

most importantly in the context of sexual education, mixed gender sports, and field trips. Another focus point is the practice of wearing a headscarf, especially if this practice is enforced in a discriminatory manner by Muslim men. The chapter concludes with an analysis of the interplay between several integration indicators and religiosity, stating that integration correlates negatively with the degree of religiosity.

The analysis of integration indicators follows the classic Esser-inspired framework of structural integration factors. The study lists indicators in the "usual four dimensions (structural, social, emotional and cultural)" of integration which are discussed with a stress on "hard integration factors": employment rate, income, vocational position, as well as language skills and acquisition.[156] Here, the stress lies clearly on indicators of social integration, such as membership in civil society organizations, contact with autochthon Germans, and openness towards bicultural marriages. Emotional integration is relatively briefly scrutinized with just two items in the questionnaire: respondents were asked if they were content with the neighborhood in which they lived and if they identified with Germany, their country of origin, or both.[157]

In a comparative perspective between Muslims and the control group,[158] the study concludes that Muslims usually display worse integration parameters than members of the control group. In general, religiosity seems to coincide with worse integration performance; this is especially true for headscarf-wearing women which are described as the "least well- integrated group":

> "It is noticeable that Muslim women with headscarf perform worse in regard to indicators of social integration. Muslim women with headscarf self-assess their German skills less often as good or very good, are less likely to be employed, have less contact with Germans, are less likely to be active in German associations, are more likely to reside in areas in which many foreigners live, seem to be less strongly emotionally attached to Germany, and are less likely to be naturalized."[159]

On the other end of the integration performance scale, non-practicing Muslims, as well as Alevites, usually display very good integration indicator values.

---

156 Bundesamt für Migration und Flüchtlinge 2009c, 207 f.

157 Ibid., 289ff.

158 The control group consisted of all participants of the MLD-telephone-interview who did not self-identify as Muslims.

159 Bundesamt für Migration und Flüchtlinge 2009c, 201f.

## Practical Relevance: Dispelling Myths

Bearing the institutional structure of commissioning the MLD in mind, it stands to reason that the prime political usefulness of the study can first and foremost be found in the context of the Islam Conference. As other authors have pointed out, the Islam Conference is an institution where the production of knowledge and the exercise of political power systematically interact.[160] Knowledge is used for the formation of a specific understanding of integration which allows the incorporation of Muslims into the exercise of political power. This argument follows the lines of reasoning developed in the context of generalized integration research and will be elaborated in respect to the context-specific features of this knowledge. In the second part, the analysis centers on a relatively disregarded, yet central aspect of the knowledge which does not address Muslims, but rather the German general public which is perceived as overly critical.

How can the knowledge on Muslims be used politically? According to the BAMF, this knowledge is primarily valued as information, following paradigmatic instrumentalist reasoning. The argument goes that the total number of Muslims is important for calculating needs of infrastructure investments most importantly in the area of integration and education. However, besides these examples of politically applicable knowledge, the study lists relatively few concrete policy recommendations.[161] In general, the need for enhanced integration policies in regard to language acquisition, schooling, and labor market participation is stressed. In this context, the study explicitly denies a demand for integration measures specifically tailored to the needs of Muslims.[162] All in all, despite the BAMF's claims, it seems that direct application of study results to political decisions is not overly important: in general, data refers to a country-wide level of analysis without regional or sub-regional divisions. It is questionable if such general data would actually help in planning infrastructure needs which are per definition tied to a specific local context.[163] The few recommendations are very uncontroversial, unspecific, and usually call for "more of the same" (such as the example of integration measures).

---

160 Hernández Aguilar 2014, Amir-Moazami 2018b, 94f., Spielhaus 2013, Engler 2014

161 Bundesamt für Migration und Flüchtlinge 2009c, p. 18

162 Ibid., p. 348

163 In fact, many federal countries have requested regional analysis of MLD data for exactly this purpose. Cp. Bundesamt für Migration und Flüchtlinge 2011c, 34ff.

To be sure, this does not mean that the BAMF's knowledge production on Muslims is not connected to a specific political use. This interest is, however, not primarily based on the needs of the administration to plan future integration policy tools. Rather, the study's political relevance seems to be mainly its discursive effects, which lie on the one hand in the introduction of a somewhat neutral, apolitical and objective narration of integration, and on the other hand, in the calming of the public debate through the introduction of a scientific stance of reason.

In the discussion of Muslims' integration successes, the question of political usefulness can be answered in terms similar to those used in the general integration research analyzed in the last chapter. Integration research contributes to a technical, apolitical understanding of integration which is conceptualized as a result of individual effort in the framework of governmental integration policies. In respect to Muslims, this notion manifests in two specific ways: first, by conceptualizing Muslims as immigrants whose integration is analyzed in the Esser-inspired framework, and second, by delegitimizing Muslim organizations. Regarding the first point, Muslims are conceptualized as migrants through the method of sample generation. Again, by way of sample gathering, only persons with a foreign sounding name were contacted; out of those, only those with a Migrant Background were included in the study. An interesting side-aspect of Muslims as foreigners can be found in the discussion of the share of Muslims among foreigners: For some nationality groups, for example Iranians, the share of Muslims in Germany is lower than the share of the population in the home country. Interestingly, this is explained solely with selection effects during migration, reasoning that religious persons are less likely to emigrate. The hypothesis that people stop practicing Islam in a non-Muslim country, out of fear of discrimination or any other explanation is left unregarded. This shows how this discussion is rooted in the concept of Muslims as migrants, and of Islam as an external, foreign phenomenon.

This does not mean that the method seems to be deliberately doctored to fabricate a connection where there is none; to the contrary, the MLD's methodology follows studiously established scientific procedures and enjoys rather high acceptance among scientists. However, it is important to note that the analysis of integration which constitutes the main framework of analysis for this study follows directly from the conceptualization of Muslims as migrants. This framework of analysis could otherwise not be employed in such a consistent manner.

As already mentioned, the analysis of integration follows the hegemonic approach inspired by Esser's assimilation theory which rendered specific deficits among Muslims, especially Muslim women wearing a headscarf. The BAMF uses these findings on the one hand in the above-mentioned sense, to calm the public debate, and on the other, to promote a technocratic, apolitical understanding of integration. Here, the argument goes that most of the publicly discussed controversies aim at deficits in social integration which can be rejected on the basis of the data. However, deficits in cultural integration (above all language acquisition) persist, a deficit which neatly fits the array of integration policy tools developed at the BAMF.

In the context of Muslim organizations, another depoliticizing effect of the integration paradigm can be discerned. The study examines the question of political representation of Muslims extensively.[164] This is conducted by asking if respondents knew about the work of the four Muslim organizations present at the Islam Conference, and if they felt represented by them. Here, although the data shows that a majority of Muslims knows at least one of the organizations, only a minority feels represented politically. This relatively low number could be explained by the fact that one of the most important Muslim organizations in Germany (Milli Görus) was not part of the questionnaire;[165] the corresponding numbers (16 percent of the respondents knew the umbrella organization) seem too low to be realistic. However, the study concludes that the data contradicts the Muslim organizations' entitlement to representation of Muslims in the framework of the Islam Conference.[166] Despite the methodological problems, this part of the survey should not be regarded as an overt manipulation of numbers with the aim to undermine the political position of Muslim organizations. It does, however, reveal a similar basic asymmetry which the Islam Conference is subject to, as Fabian Engler observed as quoted in the introduction to this chapter: Muslims are put under scrutiny and are required to testify, while the other side – the Germans, the majority society, the state – examines and evaluates. This construction of the DIK is the reason why Muslim organizations are scrutinized for their political mandate; in this situation, the BAMF's intention to contribute to an apolitical, neutral produc-

164 Bundesamt für Migration und Flüchtlinge 2009c, 173ff.

165 Due to allegations of political extremism. Milli Görüs participated in the Islam Conference indirectly as member of an umbrella organization.

166 Bundesamt für Migration und Flüchtlinge 2009c, p. 17

tion of knowledge supports this asymmetric construction which is, of course, the result of a political process.

Besides integration, the other core motif of political usefulness is the desire to "dispel myths" about Muslims. This strategy is analysis by scientific method of what are perceived as irrational, alarmist statements about Muslims in order to dismiss them as untrue or at least exaggerated. This notion can be considered rather typical for knowledge production on migration in general: in the 1960s, the Federal Agency for Work Placement reacted to negative press statements about the large family sizes of "Guest Workers" by stating that "only 75 Guest Workers have 10 children or more."[167] In her analysis of knowledge production in migration bureaucracies, Christina Boswell reports of a very similar concept in the European Union Commission:

> "To dispel the myths, to keep saying that the facts are this, the numbers of migrants are these, and these are the sorts of activities that we are engaged in, and you need to do all these things if you want to have a proper policy, an effective policy."[168]

In the context of "Muslim Life in Germany", as a BAMF researcher stated, a good part of the MLD questionnaire was dedicated to the aim of "dispelling myths." Public debate was analyzed for statements about Muslims which would be turned into objectified information by way of a representative study:

> "These are questions which moved the public debate and as a consequence, the policy-makers as well. So, they become part of the questionnaire and will be quantified by socio-scientific methods. This does not work with every question, but if it does work, there is objectified information for practical application. Because otherwise, in the discourse, there are only opinions which fly back and forth."[169]

---

167 Bundesanstalt für Arbeit 1965, p. 8

168 Boswell 2009b, p. 201

169 "Das sind Fragen, die die öffentliche Debatte und natürlich dann auch die Policymaker bewegt haben. [...] . Also, kommt das in den Fragebogen und wird versucht sozialwissenschaftlich auszumessen. [...] Das funktioniert nicht bei jeder Frage, aber dann, soweit es eben geht, hier eine objektivierte Information zur Praxis zu geben. Weil sonst, das, was im Diskurs läuft, sind ja Meinungen, [...] die hin und her fliegen."(Interview with a BAMF reseracher, September 2015)

This research strategy is most directly visible in the survey of Muslim "religiosity and religious practice." As already mentioned, the chapter seems at first glance like a loose collection of unrelated topics, some of which are not even directly connected to religiosity in a strict sense, such as the discussion of attending certain education programs at schools. However, the chapter is thoroughly based on references to controversial issues surrounding Muslims in public debate. Most sub-chapters are introduced with references to these debates, such as an allegedly massive refusal of education programs (mixed-gender sports and swimming classes, sex education, and school excursions) particularly by Muslim girls and their parents, as well as a negative connotation of headscarf-wearing as an expression of female suppression. The exact same topics were the subject of a 2006 expert study issued by the BAMF which presented a gloomy image of a mass refusal by Muslim students, particularly girls, to attend these types of educational classes and trips, a fact presented as an indicator of the negative influence of the presence of Muslims on integration processes, and ultimately, social cohesion.[170] Somewhat ironically, the BAMF's expertise stirred up some of the negative media attention to Muslims that the MLD study intends to dispel some years later. This is done by structuring the survey questions precisely according to the topics and, somewhat subtly, the allegations as well: in the case of education, participation in the above-mentioned items is polled as well as, in case of refusal, if religious reservations were the cause. In a similar fashion, the study examines the reasons for wearing the headscarf, which can be broadly categorized as either autonomous (for religious reasons, to self-identify as a Muslim, etc.) or as a result of external, especially male pressure (husband's/family's expectations, to be protected against harassment, etc.). In all cases, the survey results are presented as rather reassuring: Muslims are not overly religious as compared to the non-Muslim reference group; refusal of participation in sex education, mixed gender sports, and field trips on the ground of religious arguments is negligibly small, and headscarves are usually worn as a result of an autonomous decision.[171]

Despite the overall positive assessment, Muslims are not completely absolved of suspicion, either. In the context of mixed-gender swimming lessons, and to a degree field trips as well, a large part of the respondents answered

170 Kelek 2006

171 Bundesamt für Migration und Flüchtlinge 2009c, 134ff.

that their school did not offer such activities. This is interpreted as a possible outcome of Muslim pressure on school management:

> "However, the study result that many pupils were not offered such courses can as well be interpreted differently. It is possible that many schools with a high share of pupils with a Migrant Background refrain from offering such courses out of experience or fear that a significant part of the pupils objects to certain education forms; or they offer forms of education which are more likely to be accepted by the parents, such as single-gender sports and swimming courses or single-day field trips." [172]

All in all, the BAMF states that the above-quoted numbers should be interpreted with some suspicion because they could hide the true significance of the problem. In a related statement, the study suggests that many religious objectors to specific course offerings refrain from identifying as such and hide their true motivation behind less controversial reasons. To support this interpretation, a somewhat alternative representation of the numbers is presented:

> "To round up the image and to accentuate the share of 'real' objectors, only those pupils are considered for whom such course offerings existed and who either participated or objected for religious or other reasons. Even if only the 'affected' pupils are regarded, a large majority of Muslims as well as non-Muslims with a Migrant Background attends these courses. However, participation among Muslim girls in swimming lessons and field trips is a problematic area, of which, after all, a share of 7 and 10 percent, respectively, fails to attend. The lesser frequency of attention in comparison to boys is statistically significant so that a gender-based unequal treatment of Muslim girls in regard to these two courses can be noticed."[173]

In this context, some features of the analysis are startling: every data point which suggests so is duly mentioned, while some data points (for example, the lower attendance rate among Muslim boys for sex education) are ignored. In a related issue, the large group of respondents who were not offered such a course by the school is explained by the subtle pressure of Muslims on school management, but neither evidence nor supporting data for this allegation is presented. As a result, the overall degree of "objectors" to mixed-gender sports

---

172 Ibid., 190f.

173 Ibid., 190f.

and swimming classes as well as sexual education is exaggerated by analyzing only a part of the data, therefore enlarging the corresponding share from a negligible 1% to a considerable 7%. It seems that in this particular question, evidence has been stretched until it confirmed an ex-ante belief.

Again, while most of the study items in this section are geared towards soothing negative statements in public debate, the BAMF seems anxious to take a not too openly optimistic stance about Muslim religiosity and its influence on integration. In this context, the main direction of "myth dispelling" is coupled with an undercurrent of deficit and problematization. This is most clearly visible in the discussion of attendance to various courses offered at school with a common denominator of Muslim sexuality.[174]

Overall, this assessment of problematic behavior completes the picture of myth dispelling: the topics considered worthy of evaluation stem almost exclusively from xenophobic allegations in public debate and revolve generally around either topical areas of sexuality (gender inequality, abnormal sexual practices) or threat (violence, terrorism, abuse of Muslim women). In this context, the BAMF seems eager to establish a voice of reason: public debate is to be influenced, and made more rational, by "delivering sound facts" and by "examining the truth." Doing this, two main techniques are applied: first, epistemic authority is installed by gathering large amounts of data and presenting it with the air of the authority of both the state official and the social scientist. Second, this data is presented in a way that makes it clear these problems are not as large as imagined, but nevertheless exist, as seen in the previous example of attendance to multiple-day field trips. The resulting narrative frames a social problem that is too small to cause alarm but too big to be ignored.

All in all, by putting the two political uses of the study – integration and myth-dispelling – together, the message conveyed in the study is not that Muslims are "normal," "not worthy of special attention," "a normal part of society", etc., but rather that problems exist and will be mitigated through prudent policy-making.

174 I owe many of the arguments in the following section to Hernández Aguilar 2018, 83 ff.

## Effects on the Knowledge: The "Gaze from Nowhere"

To recapitulate the analysis so far, the two larger topical parts of the study, religion/religious practice and integration, have been analyzed in terms of their usefulness for governmental purposes. Two main results of these concepts are, on the one hand – in line with generalized integration research – a tendency to transform the subject into a task for the administration and thus depoliticize it; on the other hand, the desire to establish "neutral", "objective" knowledge akin to Haraway's notion of "the gaze from nowhere".[175] This knowledge is used to dispel myths about Muslims and thus silence what is perceived as irrational, xenophobic criticism. In the following paragraphs, the effects of these two practical uses of the knowledge will be discussed.

In the context of Muslim integration, the theoretical structure as laid out in the last chapter, as well as some of its implications, are visible as well but with specific modifications. As analyzed in Chapter 3, the Research Group's strategy to retreat from political statements into a position of scientific neutrality correlates with the perceived potential for political polarization of a given topic. In the context of Muslims, this correlation is highly visible, and not especially surprising given the high potential for political controversy connected to the issue. In the study, the construction of scientific neutrality can be analyzed quite well for this reason. To carve out this perspective, it is worth considering the research aim of the study in general: an often-cited motif in the BAMF's studies on Muslims is the lack of knowledge about them. "There is a lack of reliable information on the Muslim population in Germany,"[176] the president of the BAMF, Albert Schmidt, states in the introduction of the MLD. Similarly, "not much is known," the BAMF states, "about mosque communities in Germany."[177] These statements do not disclose the important information of who exactly does not know and what kind of knowledge is actually missing. However, if the research design is read as an attempt to answer these questions, two aspects are of interest in this context: first, as already mentioned, the research design relies primarily on quantitative methods. This is connected with the creation of a data set on Muslims which meets the require-

175 Haraway 1988

176 Bundesamt für Migration und Flüchtlinge 2009c, p. 4

177 Bundesamt für Migration und Flüchtlinge 2012e, p. 15

ments of statistical representativeness.[178] The lack of knowledge seems to be most pressing in regard to statistical, administrative knowledge. Second, the research design introduces a perspective of the social scientist which partly answers the question of who does not know: it is not, for example, the Islamic scholar interested in theological questions of Muslim religiosity in the context of migration; or the security apparatus assessing potential threats of terrorism or radicalization. Rather, the perspective of the social scientist emerges who charts the Muslim population in Germany by way of a representative survey. This fact is presented as the key difference between the MLD and all other socio-economic studies on Muslims, which were usually complied on municipal or regional levels.[179]

Conceptually, however, the notion of Muslim religiosity remains strangely vague: as already mentioned, the relevant questionnaire items mainly consist of self-identification as Muslim, the observance of selected ritual practices, and a discussion of "anti-Muslim" accusations. The focus lies clearly on externally visible features of the religion and does not touch upon questions of faith and belief systems; there is no desire to map out the inner constitution of German Islam and its different theological, ethnic, or socio-economic layers. At the same time, the prevalent security perspective on Islam is absent as well, visible for example in the omission of topics such as radicalization or extremism in the questionnaire.[180] In effect, the research position of the BAMF is by no means neutral, or objective; rather, it seems that the BAMF's research focus on Muslim religiosity is that of an outsider. This can be seen in the fact that the analysis does not discuss basic terms and categories connected to the religion such as faith, specific beliefs and their inner logic, and so on. In MLD, Religion appears primarily through the socio-scientific view: items which can be checked in a questionnaire, especially those that are visible from the outside, are stressed. In other words, the logic of "avoiding political controversies" and "scientific neutrality" requires establishing a research perspective on Muslim religion *from the outside*. Quite tellingly, literature ref-

---

178 This stress on quantitative methodology can also be seen in the fact that the relatively modest qualitative items in the original research design – a comparison to other European countries, for example – have apparently been eliminated later on. Cp. Bundesamt für Migration und Flüchtlinge 2009b, p. 36

179 Bundesamt für Migration und Flüchtlinge 2009c, p. 4

180 Schepelern Johansen and Spielhaus 2018, p. 147

erences stem almost exclusively from social sciences methodology and refer to the question how to measure religion with quantitative methods.

This outsider view is of course not a neutral one, but rather a nationalized German perspective. This fits neatly to the conceptualization of Muslims as "foreign". This is inscribed into the method of sample generation as discussed above. While this finding is discussed in respect to the selectivity processes of migration, it also confirms the hypothesis that the concept of Muslims is based on certain ex-ante assumptions, the most important of which is that the typical Muslim is a migrant from a non-EU member state. Schepelern Johansen and Spielhaus (2018) connect this framework of analysis to an "epistemic inertia" of the field, thanks to which time and again outdated concepts of belonging constantly reemerge:

> "Although the sample method [...] might be founded in pragmatic reasons and the excluded persons may even not be statistically significant, these studies display evidence about the [...] epistemic hegemonies of the field. Again, the national state plays the dominant role as a prominent background, by equating Migrant Background with Being Muslim and by maintaining the intermixing of the categories 'Muslim' and 'Migrant'."[181]

In "Muslim Life in Germany" it is reasonable to assume that this epistemic inertia is a direct result of both the conditions of knowledge production and the political implications of the sensitive topic. For a governmental researcher, a recurrence to the national state as a point of reference is not a methodological flaw but rather a prerequisite. This is visible in the geographical level of the research – all of Germany instead of single towns or regions – which is presented as a decisive improvement over other, small scale studies. The conceptualization of Muslims as migrants follows from a construction of Germans as an ethnicity, which refers to a specific idea of the nation state as well. In addition to that, the merge of the categories Muslim and migrant is a direct prerequisite to apply a framework of analysis of integration, which in fact only makes sense if it is applied to foreigners. This is a result of the intellectual tradition of the Research Group whose expertise in this field predisposed the application of an according framework of analysis. The distinction of religious acts into what can be seen from outside (attending service at the Mosque) and what not (degree of religiosity) correlates to the overall perspective of integration: Muslim religiosity is not analyzed in its capacity

181 Schepelern Johansen and Spielhaus 2018, pp. 139–140

as a religion, its properties as a group identity, or its political implications. Rather, religiosity is conceptualized as a resource (or challenge) in regard to integration.

Of course, one could put forward an argument of triviality here: after all, an outsider-perspective on religion and a stress on integration does not seem overly surprising when originating from the BAMF. However, it does constitute rather an exception than a rule among governmental documents on Islam at the time, which is most often discussed from a public security or at least deficit-oriented point of view which stresses the presence of Muslims as a threat to security or to social cohesion, respectively.[182] Aspects of public security are curiously absent in the analysis part of the "Muslim in Germany" study; neither does the questionnaire contain items such as radical/fundamentalist world-views, nor is the analysis specifically geared towards this particular subject.[183] The rare references to the discursive link between Islam and fundamentalism are discussed within the framework of myth-dispelling as a part of exaggerated allegations.[184] This is in a way remarkable since it is untypical for German Islam policy in general, and the Islam Conference in particular.[185] At the same time, the above-described "outsider perspective" and the political use of calming the public debate is in some respects almost the reverse angle of view of an earlier BAMF study on school attendance of Muslim children in sex-ed, mixed-gender sports, and field trips: the perspective is that of an insider, or at least an expert on Islam; the methodology is qualitative and does not claim representativeness; and the results of the study are rather alarming.[186]

Following Schiffauer's analysis of knowledge production in security bureaucracies, the integration perspective in the MLD can be connected to an alternative governmentality of Islam originating in the Ministry of the Interior which attempts to tap the organizational resources of the Islamic community in Germany for integration policy-making. This approach, however, is

182 Müller 2018, 189f.

183 Questions of radicalization are briefly discussed in the literature review of the study, however, rather in the context of deviance and criminality than terrorism. Cp. Bundesamt für Migration und Flüchtlinge 2009c, p. 30

184 Cp. for example Bundesamt für Migration und Flüchtlinge 2009c, p. 193 in the context of the practice of wearing a headscarf.

185 Müller 2018, 189f.

186 Kelek 2006

at odds with the otherwise hegemonic "security knowledge" inside the Ministry which conceptualizes Islam primarily as a threat to public security.[187]

Coming back to the discussion of research perspective, it is not the aim of this analysis to make a judgment if the BAMF's angle of research is more useful or objective than any other discussed here. Rather, it is important to note the inherent dilemma in producing politically relevant knowledge which is at the same time objective and neutral. This dilemma is well visible in "Muslim Life in Germany" and can be analyzed in the difficulties of introducing a focus on integration in equidistance to both security knowledge of the administration and "insider knowledge" of Islamic faith. In this battle of ideas, the BAMF's research report shapes a specific understanding of Islam in two ways: by excluding security-related issues from the analysis, and introducing a perspective of integration, the potential of Islam as a resource to integration is underlined; at the same time, the analysis is visibly geared towards presumed integration deficits of Muslims which are presented as opportunities for integration policy planning.

## Conclusion

To summarize, the BAMF's study on Muslims – one of the largest studies the Research Group has ever conducted – is a typical representative of governmental integration research at first glance. The application of the classic Esser-inspired framework of analysis of integration, the focus on quantitative methodology, and the production of a research perspective of scientific neutrality all seem rather common. However, Muslims are a fairly exceptional research object, especially thanks to the increased public attention to the topic and, regarding the institutional structure of knowledge production, the involvement of the Islam Conference as a study contractor. As a result, a specific understanding of the Islam is formed which conceptualizes Muslims as migrants; this knowledge hides its context-specific perspective behind a seemingly "neutral" and "objective" narrative.

As the analysis has shown, the knowledge on Muslims is relevant for the political process in the most part in its capacity to calm public debate. The reasoning behind the strategy is that racism is based, in the end, on wrong information, which can be overcome by providing better, more accurate numbers. However, by structuring the questionnaire in answer to racist discourses, the

187 Schiffauer 2018

resulting knowledge is inherently structured by them and makes only sense within their confines. The swimming-pool discussion in MLD and elsewhere is illustrative in this context: The number of girls failing to attend swimming lessons for religious reasons is, per se, not important statistical information; it becomes only relevant data in its capacity of myth-dispelling, or in other words, in the context of xenophobic discourse. The same holds true for large parts of the survey on Muslim religiosity: the questions about attendance to sex education, mixed gender sports, and other items render insignificantly low rates of refusal on religious grounds; the questionnaire items on wearing the headscarf similarly dismiss notions of external pressure and coercion for this practice. The subsequent knowledge, however, does not contribute to a better understanding of the religion's role in public, or vis-a-vis the education system; it is a bound knowledge specifically situated in the contemporary public discourse. By taking up myths, despite the usually dismissive result of the study, the object of scrutiny is singled out for examination and therefore marked as something specific, defiant, or in need of explanation. In other words, somewhat ironically, the technique of dispelling myths with scientific means contributes to the construction of the very same myths, and the construction of a specific group of others.

## Migration Potential

In the last chapters, different knowledge/policy complexes from the main research fields of the BAMF – integration research, migration reporting, and Muslims – have been analyzed in regard to the practical applicability and the epistemic features of governmental knowledge. In each of these cases, the dynamic development of a situation-specific governmental perspective can be discerned, which changes according to practical relevance deliberations. As evident from the analysis of the Migration Reports and integration research, both publication forms have a distinct style of avoiding direct references to theoretical concepts and refer instead in many instances to common sensical knowledge generated and maintained by the state administration. On the other hand, this knowledge refers implicitly to scientific theory as well, albeit usually briefly and altered according to "practical relevance" deliberations. The differences between scientific theory and what is considered "practically relevant" constitute a core characteristic of governmental knowledge in the analysis so far.

In this respect, the notion of migration potential is an exception to this rule: the term is relatively clearly identified as a theoretical concept in the academic sense of the term, with the usual features such as scientific definitions, references to academic literature, an operationalization concept and a distinct methodology of analysis. Migration potential features almost from the beginning among the research projects at the BAMF.[188] Over the years, several research projects deal explicitly with this research topic. Among those are research reports with analyses of migration potentials in Africa (published in 2009), the Commonwealth of Independent States (2012), and India (2013).[189]

---

188 Cp. Bundesamt für Migration und Flüchtlinge 2009b

189 Bundesamt für Migration und Flüchtlinge 2009d, Bundesamt für Migration und Flüchtlinge 2012a, and Bundesamt für Migration und Flüchtlinge 2015b. The migration potential of African migration is the topic of a "publication series" anthology as well (Bundesamt für Migration und Flüchtlinge 2011g) which discusses the term in a more diversified way, including for example a discussion of effects of migration on the sending countries, or multi-faceted analyses of integration. These texts, however, are not part of the "official body of knowledge" for two reasons: first, they represent only the author's opinion, and second, they are not subject to the same production logic as the knowledge generated at the Research Group. This can somewhat be supported by the fact that the BAMF's contribution to the analysis of migration potential (Schmid

Furthermore, the concept appears in several working papers and Migration Reports in different topical and regional contexts.

All in all, both the exceptionally high profile as a theoretical concept, its proliferation especially in regional migration studies, and the long period of time in use are indicators justifying a detailed analysis of the concept.

In the next paragraphs, the historical development of the theoretical concept will be briefly outlined before its practical relevance is discussed. The chapter concludes with an analysis of the epistemic features of the knowledge, focusing especially on potential bias sources arising from the practical applicability requirement of knowledge generation.

## Migration Potential and Potential of Migration

As already mentioned, the concept of migration potential has a long tradition among the research projects at the BAMF and features among the earliest research projects practically from the beginning of the research activities in 2005. The development of the concept of migration potential and its potential use in the political system is described by Research Group staff as a somewhat complementary understanding of two different aspects of the term: "potential of migrants" and "potential future migration":

> "We address two different aspects of migration potential: both the aspect of the potential of migrants, because it is important to focus on what they bring with them, what can be useful. That is one aspect, the other is the question which is directed to us time and again: 'who will come' and above all, 'how many are going to come?' Accordingly, we have tried to analyze different world regions to find out what determines migration and in how far can it be anticipated or at least discussed how it develops further. We have relatively quickly abandoned [the idea] to compute a prognosis because of the data quality in other states. Instead, we discuss the factors which influence migration from these regions."[190]

2011) is a shortened version of the above-quoted research report with an unaltered theoretic and methodological framework.

190 "Migrationspotenzial [...] [haben wir] aufgegriffen unter zwei Aspekten: Sowohl unter dem Potential des Migranten, [also man] sich anschauen muss, was die mitbringen, was auch nutzbar ist. [...] Das ist die eine Richtung, [...] die andere ist die Frage, die immer wieder an [uns] gerichtet wird: 'Wer kommt denn da eigentlich.' Und vor allen Dingen, 'wie viele werden dann noch kommen [...]'? [...] Dementsprechend haben wir

This understanding is the result of a development over the years, during which the term underwent considerable changes in respect to the research interest, the methods of analysis, and in some respects the theoretical references. These changes will be retraced in the next paragraphs.

In 2006, the Migration Report mentions migration potential for the first time. In this report, the term is discussed in the context of a prognosis of future migration to Germany. Future migration is believed to be dominated by the same regions of origin as in the past, namely Southern Europe and Turkey, which is why the Migration Report focuses on these regions. In respect to these countries, the analysis of migration potential centered especially on demographic data to predict future migration streams. While the report notes that demographic differences alone are not sufficient to predict migration, demographic data is used nonetheless since it is the only data available for long time horizons. In this context, the report features the only concrete prognosis of future migration the Research Group has ever published concerning migration potential: The BAMF estimates the long-term average of future immigration to 100,000 to 200,000 persons; the numbers are quoted from a similar prognosis of the Federal Statistical Office.[191]

Visually, this prognosis is quite interesting: historically, migration data is characterized by stark yearly fluctuation instead of smooth long-term developments. The prognosis of migration potential translates this history into an orderly, narrow corridor of potential future migrations. Methodically, the corridor has been simply calculated from the long-term average between 1995 and 2005. The BAMF states that there is no better method available since there is no distinct trend visible in the past development which could be used to extrapolate a somewhat more refined trend. The BAMF states that the corridor

uns an Analysen von verschiedenen Weltregionen versucht, um zu schauen, was determiniert dort jeweils Migration und inwieweit, kann man eventuell voraussehen oder zumindest diskutieren, wie sich so was weiterentwickelt. Wir haben uns schnell davon verabschiedet, [...] eine Prognose [zu] errechnen, aufgrund der Datenqualitäten, die es in anderen Staaten gibt [...]. Sondern wir diskutieren die Faktoren, die Migration aus diesen Regionen beeinflussen."(Interview with a BAMF researcher, 2015)

191 Cp. Messerschmidt 2017, p. 319. According to the author, demographic prognoses construct a notion of an inevitable "natural force" by reframing past political decisions into questions of population. This inevitability, as well as the discursive power of the Federal Statistical Office's population prognosis, is to a degree reproduced by the BAMF in this context of prognosis of future migration movements.

does not represent a concrete prognosis of future migration movements; it does, however, display the long-term average of expected future migrations.

The theoretical understanding of migration potential is elaborated further in two large research projects between 2009 and 2012 that study migration potential from Africa and former member states of the Soviet Union (CIS). According to the yearly report of the Research Group, these two projects aim at developing an "estimation method of future migration movements." This is conducted by an "analysis of circumstances in selected regions of origin and immigration" and a development of theory as well since "common assumptions are to be specifically evaluated and developed."[192]

Especially in regard to theoretical development, the studies make exceptional references to academic literature. The makeshift analysis in the Migration Report and the prognosis of future migration which exclusively referred to demographic differences as a point of departure is substantiated with a theoretical framework.

This framework elaborates not only on the understanding of migration potential, but explains the basic mechanisms of migration as well, applying a neoclassic theoretical model which basically consists of a rational choice micro analysis and a push-pull factor macro analysis. According to this theory, migration is the result of a single, unidirectional, individual decision conceptualized as a rational choice of utilization maximization; in simple words, migration is preceded by a comparison of advantages against disadvantages plus transaction costs. This decision-making process takes place in a macro-framework of analysis – commonly known as push-pull factor analysis – of differences between regions:

> "Every migration movement is preceded by an evaluation of advantages and disadvantages between emigrating and staying, as well as between the attractiveness of a target region and the size of obstacles which have to be overcome on the way. If a positive decision is taken, the person will emigrate."[193]

Following this reasoning, the migration decision is mostly influenced by push and pull factors between the migrant's home country and a given destination. Based on this idea, the notion of migration potential is introduced. This potential is defined as "[the sum of] all potential migrants in a region, which do

192 Bundesamt für Migration und Flüchtlinge 2009b, 23f.

193 Bundesamt für Migration und Flüchtlinge 2009d, p. 23

not see a perspective of fulfilling life goals locally and thus want to emigrate to more prosperous regions at an unspecified [...] point of time."[194]

Both the CIS and Africa research reports list factors and data from four main areas: demography, economy, politics, and natural environment.[195] These areas are different to the earlier prognosis model which focused exclusively on demographic differences; however, demographic differences are still clearly the centerpiece of the argument, since many factors from other areas (especially the economy and environment) are framed as factors dependent on population data. In this context, the change of argumentation is somewhat appalling: the earlier prognosis model approaches the demographic data tentatively and defensively by claiming that it is only used in the absence of better numbers. According to this argument, demographic data alone is insufficient to predict future migrations, but is used nevertheless since it is relatively easy to estimate over a long period of time, unlike other socioeconomic data.[196] In contrast to this, the new framework of analysis centers precisely on the demographic differences, which are portrayed as the core factor influencing economic, environmental, and to a degree political aspects of migration potential as well. This focus is clearly visible in regard to the depth of analysis as well: demographic data (and to a lesser degree economic data) is discussed thoroughly, while political and ecological factors are swept over rather briefly; together with the relative length of the respective chapters, the overall impression is created that migration potential depends mostly on demographic and economic disparities.

With this change of methodology, a change of the research goal has been introduced as well. The aim of a concrete prognosis has been given up since the first draft; unlike the Migration Report in 2006, no concrete numbers of future migrations are released. Instead, the notion of migration potential is redefined in the following manner:

> "Migration potential [is] not an exactly calculable number, but rather a 'collective phenomenon' which results from existing tensions and differences in development. Research on migration potential cannot be understood as an instrument of prognosis of concrete migrations, but rather aims at the contrasting of social spaces with different life chances. This reveals a migration

---

194 Ibid., p. 23

195 Ibid., 34ff.

196 See critically, Messerschmidt 2017, p. 353

> potential and can be connected to current and future problems, which have to be taken into account by political actors."[197]

As a result, the analysis focuses almost exclusively on the display of predominantly demographic and economic factors which support the ex-ante assumption of a high migration potential. In the example of the Africa study, demographic discrepancies are discussed in relative length, referring mostly to a UN prognosis of demographic data until 2050. "African" populations are typically represented by what the report calls "least developed states," referring mostly to states in Central and Eastern Africa. These regions are characterized by high birth and mortality rates, a high population growth, and a low average age of the population.

Likewise, economic data is presented in a manner that leaves the impression that practically the whole continent suffers from poverty, unemployment, and low income. Concerning the categories of environment and politics, the author states that it is impossible to provide numerical data. In the absence of these, several indices (Human Development Index, Corruption Index, Failed State Index) are presented. Again, a negative image of the African continent is drawn which seems to consist of corrupt bureaucracies, regimes with human rights violations, and press censorship. In the context of environmental factors, likewise negative effects of climate change and natural catastrophes (volcano eruptions, etc.) are presented as indicators for a growing migration potential. Based on these characteristics, a dichotomist picture is drawn: underdeveloped Africa (and Eastern Europe) on the one side, industrialized, rich and developed Europe on the other. This difference is the most important outcome of the analysis since, as the BAMF concludes, this differential "in the last instance creates the migration potential."[198]

In more recent studies on migration potential, the term has been given yet another meaning and context.[199] The studies are not regionally confined, but examine migration in connection with newly created immigration possibilities for temporary work migration. In a study on migration from Romania and Bulgaria, the future potential of migration after the two country's accession to the Schengen Area is discussed. By and large, the study uses a similar theoretical framework: migration is explained by economic and demographic

---

197 Bundesamt für Migration und Flüchtlinge 2009d, p. 199

198 Ibid., p. 34

199 Cp. for example Bundesamt für Migration und Flüchtlinge 2014e, Bundesamt für Migration und Flüchtlinge 2015b,

differences analyzed in a push-pull framework. However, two methodological differences in the analysis, as compared to the above-mentioned studies, are visible. First, the framework of analysis is less rigidly structured according to theoretical deliberations and more data-driven. While some theoretical assumptions – such as push-pull analysis – are still upheld, the major part of the study is dedicated to the analysis of empirical data. Here, different data is presented for every country individually, including both state and non-state academic sources. The style of analysis in this context constitutes the second major difference to the Africa/CIS studies: instead of deriving an ever "growing migration potential" from every piece of information presented, the Bulgaria/Romania study presents differentiated and sometimes even inconclusive data. All in all, conclusions which are drawn can be characterized as rather tentative and mid-range: instead of repeating an ex-ante assumption of a growing migration potential, the study differentiates between different forms of migration – some of which might be growing, some of which decreasing, in the future. For example, the study connects a growing influx of low-skilled Romanian workers from Spain and Italy due to the economic recession in these two countries, but assumes that this immigration will cease once the economic situation ameliorates.[200] Likewise, migration is qualified in terms of the length of the migration project, and pendular migration forms; in this context, the study concludes that a large share of migrants will stay for a short period before returning to their home countries. Instead of conceptualizing migration as a unidirectional, once-and-for all decision, the analysis includes pendular, short-term, and otherwise atypical forms of migration which are for the most part disregarded in the African and CIS migration studies. All in all, the study concludes that the migration potential is slowly growing, but assesses this fact rather positively since most migrants find employment either as skilled or unskilled workers and can therefore be considered useful.[201]

In short, the development of the term "migration potential" can be described in three stages, from a prognosis model in the 2006 Migration Report to a relatively elaborate theoretical model in the studies on migration potential from Eastern Europe and Africa to an analysis instrument of intra-EU migration movements. If this history of development is compared to the introductory statement of a connotation of both "future migration" and "potential

200 Bundesamt für Migration und Flüchtlinge 2014e, p. 138
201 Ibid., p. 150

of migrants," it seems clear that this connotation is not quite complementary, but rather segmented: in the context of EU migration and the labor market, the potential of migrants is underlined, while in the context of Eastern Europe and Africa, the (dangerous) potential of future migration is stressed.

## Practical Relevance: Ex-Post Legitimization

In its original design, the concept of migration potential aimed at one of the most sought-after scientific inputs into politics: the prognosis of future migration streams.[202] The correct prognosis of migration can therefore be regarded as a highly relevant, practical application of knowledge, which lies at the initial impetus of the according research work. However, the concept of migration potential revealed – especially in regard to its prognosis capability – some weaknesses. In the case of the Eastern Enlargement of the European Union, migration potential studies have been conducted to predict future immigration movements. In an analysis of more than 20 studies on future migration movements from Eastern Europe to Germany, Brücker/Baas (2010) describe typical methodological and empirical shortcomings of migration potential studies.[203] Methodically, the studies used either surveys among potential migrants or (more often) prediction models based on mostly economic and demographic variables. In most cases, a high emigration pressure was presumed, which led to a long-term prediction of about 3 to 5 percent of the population emigrating, which would translate to about 200,000 to 450,000 emigrants annually. Very similar to the BAMF Africa Study, in most cases an "immigration pressure" (from the perspective of Germany) was presumed, based on the perceived differences in economic development, income, unemployment, purchasing power, and other factors between Germany and Eastern Europe. While Brücker/Baas state that it is "not absolutely certain that all predictions have been disproved by actual developments,"[204] it seems clear that the actually realized migration potential lies definitively at the lower end of the prediction corridor. Additionally, this relative precision has only been achieved by compiling aggregate numbers. Original studies which usually focused on single countries of destination numbers were much more likely to

202 "'Wer kommt denn da eigentlich.' Und vor allen Dingen, 'wie viele werden dann noch kommen [...]'?"(Interview with a BAMF Researcher, 2015

203 Brücker and Bass 2010, 31ff.

204 Ibid., p. 31

overestimate the migration potential and predict a too high potential for future migrations.[205] All in all, it seems that the problem with migration prognosis is not just a matter of data quality, but rather a fundamental problem which can be connected to the high degree of uncertainty of future migrations, as well as to the fact that most assumptions of neo-classic migration theory have been disproved by more recent research; the method of deriving an unfailing migration potential on the basis of a push-pull factor analysis seems rather questionable.[206]

The abandonment of the development of a prognosis instrument can be regarded rather consequential of the various methodological drawbacks of the concept; however, knowledge production on the topic has not ceased to exist but rather has taken new forms. In order to analyze the political usefulness of the reformed concept of migration potential, it is worth reflecting shortly on the changes between the 2006 Migration Report and the later research reports on African and CIS migration: in this respect, the analytical focus shifted from the main regions of origin to rather unimportant sending regions. In this combination, this selection seems rather odd at first glance: neither region is especially important in terms of origin of migration streams. In the example of Africa, the study notes that merely 3.7% of foreign nationals in Germany possess a passport of one of the African states, which amounts to 0.3% of the total population.[207] The same can be said about unregistered migration from the continent, which likewise does not play a significant role in terms of volume and impact of inflows.[208] In both Eastern European and African migration, circular, intra-regional migrations outweigh migration to Europe by far, a fact which is not easy to reconcile with neoclassic assumptions of utility maximizing.[209] The history of migration and resulting differences in the legal framework are other factors which hinder the comparability between the target regions of the studies: in the case of Africa, migration is organized either in the asylum process or via family reunification; in the case of Eastern Europe, "Ethnic German" immigration plays a dominant role.[210] Not only is it questionable if these fundamentally different migration systems can

205 Bundesamt für Migration und Flüchtlinge 2014e, 54f.

206 For a detailed critique, see Massey et al. 1998, 10ff.

207 Bundesamt für Migration und Flüchtlinge 2009d, p. 145

208 Vogel and Aßner 2012, p. 35

209 Cp. Marfaing 2011

210 Bundesamt für Migration und Flüchtlinge 2009d, p. 144

be analyzed within the same framework of analysis, but also in both cases the application of neoclassic theory seems rather unfit. Push-Pull factor analysis was initially developed to explain domestic migration, which assumes the absence of transaction costs (other than geographic) and therefore systematically disregards the effects of unequal granting (or restriction) of mobility rights as present in the case of both African and CIS migration.[211]

In short, the analysis of Eastern European and African migration with the concept of migration potential seems off for two reasons: neither region is especially important in terms of inflow, nor does migration potential as a theoretical concept adequately describe the actual dynamics of migration. This leads to the question why this particular approach has been selected, and why these two regions represent an object of interest to knowledge production. In relation to the latter question, the BAMF explains that the two regions were selected since they represent the two main regions covered by the EU's Global Approach to Migration and Mobility (GAMM).[212] This assessment of the Research Group follows the political practice of the European Union to balance migration policy initiatives between Eastern Europe and Africa, which justifies in a way the shift in knowledge production despite the lack of actual migration movements to Germany.[213] In this context, the shift of knowledge production towards two less important world regions (in terms of inflow to Germany) becomes clearer, since it can becontextualized in the marked tendency of EU migration politics towards technical measurements of border surveillance and combating irregular migration. In the context of security-related aspects of irregular and transit migration, the two regions under scrutiny here feature highly on the priority list of EU-politics, as can be seen, for example, in the fact that "mobility partnerships" were built predominantly with states from either region.[214] In this context, it stands to reason that not only the geographic focus, but also the framework of analysis was chosen to support the policy: migration from Africa and Eastern Europe fits well the ex-ante assumption of a high migration potential despite the lack of

211 Bundesamt für Migration und Flüchtlinge 2009b, 23f.

212 Bundesamt für Migration und Flüchtlinge 2014d, p. 16. For the GAMM, see European Commission 2011

213 Angenendt 2012, p. 20 As of 2014, Bilateral Agreements have been concluded with Moldova (2008), Kap Verde (2008), Georgia (2009), Armenia (2011), Morocco (2013) and Aerbaidzan (2013) (European Commission 2014, 2f.).

214 Cp. Kratzer 2018b

corresponding migration movements. This potential is further exaggerated by presenting a biased selection of data which almost inevitably points in the same direction: the migration potential is high, and growing, despite the relatively low numbers of actual immigration.

Taken together, it seems clear that with the geographical shift to Africa and the CIS, the practical use of the concept no longer primarily lies in its prognosis potential. Instead of predicting future migration movements, migration potential analysis is employed as an ex-post legitimization of political decisions, namely the European strategy of closing external borders while maintaining a relatively high degree of freedom of movement within its territory. In this context, the theoretical understanding of migration potential is constructed in a way that it cannot be verified against empirical data, and data is presented in a unidirectional way to support the hypothesis. According to neoclassical theory, a high migration potential can be deduced from a differential in life circumstances, economic and political development, and so forth. The legal and technical barriers at the external EU border – conceptualized in this theoretical model as transaction costs – are the only restraining factor standing in the way of actually realizing this steadily growing migration potential. On the other hand, the Research Group's research on intra-EU movements – such as migration from Romania and Bulgaria – renders less alarming facts: the economic advantages of migration are underlined, and the overall outlook is positive. This notion is especially evident in the study on migration from Romania and Bulgaria which is evaluated rather positively and in any case not as a threat to social and economic order; in this context, migration potential describes the potential *of* migrants in terms of their capacities as laborers and because of their favorable demographic features. Here, migration potential does not signify a danger, but an asset for economic growth. It is interesting that the very same features serve in the case of African migration as arguments for constructing a dangerous migration potential through uncontrollable immigration pressure.

To summarize, the evolution of the migration potential concept can be explained in terms of its usefulness in a political sense: at the beginning of the research activities, prognosis of future migration constituted the most sought-after political use of expertise. However, these early concepts of a prognosis instrument were abandoned in exchange for a model of push-pull analysis whose practical use lies in legitimizing policy: by drawing an alarming picture of the migration potential of unwanted migration streams, such as from Africa and Eastern Europe, restrictive measures of border control and surveillance are legitimized even though actual migration movements are not overly significant in terms of their volume.

At the same time, the very same features of intra-EU migrations are positively evaluated, thus contradicting to a degree the theoretical arguments developed before.

All in all, the concept of migration potential can be useful to support different positive or negative aspects of migration along the lines of EU migration policy: in the context of migration from Africa and Eastern Europe, the concept of migration potential is a useful theoretical foundation to justify policies of closure and surveillance; in the context of intra-EU migration, the same concept is geared towards a notion of the potential usefulness of migrants.

## Effects on the Knowledge: "Fuzzy Logic"

In the last paragraphs, the political usefulness of migration potential has been analyzed. By altering the notion from a prognosis instrument to a push-pull analysis focusing primarily on differences, the discursive role of this knowledge lies rather in legitimizing ex-post political decisions than in informing them. While this usefulness can in fact be regarded a success of the conceptualization – the Africa study is among the most popular research papers, according to the BAMF[215] – it stands to reason that this remarkable career of the notion came at a cost in terms of its epistemic quality.

The change of the concept from a prognosis to a legitimizing instrument was accompanied with a shift of theoretic foundation and target regions of the analysis. In regard to theory, the neoclassic framework of analysis, with a clear focus on push-pull factor analysis, was introduced, replacing the linear extrapolation of demographic data of the 2006 Migration Report. The relatively one-sided structure of argumentation – intra-EU migration viewed as positive, whereas the dangers of *potential* migration from Africa and Eastern Europe are underlined – leads to the conclusion that the knowledge is not used for the stated purpose of forecasting migration movements but rather for the ex-post legitimization of given political decisions, namely the EU Global Approach to Migration and Mobility (GAMM). In the next section, these factors are discussed in regard to their influence on the epistemic quality of the gen-

215 Bundesamt für Migration und Flüchtlinge 2015a, p. 20

erated knowledge in two respects: first, the theoretical quality of the concept, and second, its quality as a prognosis instrument.

In regard to the evolution of what migration potential actually signifies, a major step was the abandonment of the use of the notion as a (however preliminary) prognosis. Both the African and CIS studies do not contain any predictions in numbers of that sort. Instead, migration potential is no longer understood as a scenario of the probable development of future migrations, but rather replaces a probabilistic scenario with possible migrations which *might* manifest in the future. Connected to this shift in research interest is a change in the application and interpretation of empirical data: the heavily theoretical approach to the phenomenon leads to a relatively monolithic and unidirectional interpretation of empirical evidence, connected with a incomplete check of theoretical assumptions against empirical data. This effect is visible for most central assumptions of the theoretical framework: the push-pull framework of analysis suggests that migration is greatest between regions with the largest differential in demographic and economic terms: in other words, between Africa and the European Union. However, this is clearly not the case: most migration movements take place within regional networks of migration and do not automatically lead to emigration to Europe as the framework of analysis might suggest. Likewise, if the assumption of a demographic pressure was true, the states with the highest birth rates and the fastest growing populations would feature among the chief sending countries within Africa, which is also not the case.[216] The same is true for forced migration movements, which likewise for the most part are regional in character; this is true even in cases when refugees flee from countries in close vicinity to Europe, such as Libya.[217] Concerning the second cornerstone of the theoretic model, demographic pressure, critical contributions are equally skeptic about the central assumption of a higher migration potential triggered by demographic discrepancies; although they concede that demographic forces influence economic development by changing the number of unemployed persons, consumers, or users of public services, the idea is refuted that this mechanism translates quasi-hydraulically into a "migration pressure" towards countries with a more favorable demographic build-up.[218] At the same time, central

216 Massey et al. 1998, p. 10

217 In 2011, about 630,000 foreign nationals fled Libya, out of which about 6% or 40,000 arrived in the European Union. Numbers quoted after Pastore 2011, 2f.

218 Massey et al. 1998, p. 11

assumptions of the rational choice model of decision-making cannot be confirmed by empirical data: migration in this context is most often not a unidirectional, once-and-for-all decision based on economic utility maximization, but most often temporary and circular in its character, as already mentioned.

To summarize, the development of the BAMF's understanding of migration potential is characterized by a double uncertainty: in the first place, it refers not only to actual migrants, but also to "potential" migrants who allegedly only wait for the next favorable opportunity to emigrate. Especially in the context of Africa, the BAMF suggests that the absence of large numbers of actual migrants is counterbalanced with a presumably large number of *potential* migrants. This potential manifests in the fact that people might not want to emigrate yet, but might as well do so in the future. With this redefinition, a quite remarkable stretching of empirical evidence is achieved: any person can be considered a potential future migrant; regional and circular migration movements can be interpreted to end in Europe in some point in the future, regardless of their actual aim and features. The fact that most migration takes place within Africa can thus be reinterpreted as proof for a growing migration potential towards the EU.[219] The discussion of environmental factors illustrates this point even more drastically: per se, it would be absurd to claim that migration between Germany and Africa was triggered by volcanic eruptions. However, by claiming that volcanic eruptions (and the lack thereof in Germany) contribute to raising the migration potential, the hypothesis that volcanic eruptions trigger immigration to Germany is turned into a possibility, however distant. This claim can furthermore not be disproved by empirical evidence, since it refers to a vague possibility instead of specifying a probability score to future events.

While it is logically impossible to disprove the migration potential framework due to its circular argumentation, it is possible to assess its epistemic quality as a prognosis model. Philipp Tetlock's quantitative analysis of expert knowledge provides a toolbox for assessing the logical construction of migration potential as a prognosis instrument. One core method of analysis is the deconstruction of expert knowledge into discrimination and calibration scores: perfect discrimination always assigns 100% possibility to events that eventually happen and 0% to those events that never happen, while perfect calibration scores assign in aggregate the correct probability to a given event

219 The European Union's border protection agency Frontex employs a similar logic in their annual risk analyses. Cp. Kratzer 2018b

(for example, an event that occurs in 60% of the cases is assigned a 60% probability). These values are often portrayed as a trade-off: to achieve good calibration, it usually pays off to assign close-to-average probabilities to events (40-60% probability), while the discrimination rate in this case is extremely poor. Such a strategy would equal a weather forecast based on the average temperature and rainfall for every day: While it is relatively unlikely that this forecast is actually true for single events, the margin of error is usually quite small; that means, the predictions are never very far off. In conclusion, the long-term averages and therefore the calibration scores of the prediction are most likely correctly predicted by that approach.

On the other hand, good discrimination scores are achieved if "bold" predictions are made. In an extreme scenario, only 100% and 0% probability scores will be assigned to a given prediction. In this strategy, the overall potential for error is higher, but so is the probability of guessing single events right – which then translates into a better discrimination, but a worse calibration score. The overall quality of a forecaster can thus be measured in a combined score of both indicators. The evolution of the migration potential model can well be captured with Tetlock's analysis: the prognosis of migration on the basis of past average numbers represents a relatively good calibration score with a discrimination score close to zero. In other words, it is relatively likely that actual yearly migration is seldom within the corridor of prognosis; but at the same time, it seems likely that long-term averages will be. If the Federal Statistical Office's migration corridor of 100,000 to 200,000 migrants is compared to past developments, this assumption is confirmed: only a minority of the yearly migration numbers of the past actually falls within the corridor, while the majority displays either higher or lower numbers.[220] Following Tetlock's approach, the prognosis capacity of the new understanding of Migration Potential cannot be regarded as a development towards higher combined calibration and discrimination scores: rather, the claim that "migration potential is high, and growing" is an even less well calibrated prognosis than the one put forward in the 2006 Migration Report, since it refers to a completely virtual concept that cannot be verified against empirical data.

In this context, it is important to note that the BAMF claims that migration potential has been developed further by elaborating its theoretical base.

---

220 If the migration prognosis corridor is projected backwards, about 30% of past data points (between 1995 and 2005) fall within 100,000 and 200,000 immigrants. Own evaluation based on Bundesamt für Migration und Flüchtlinge 2007a, p. 218

The argument goes that by doing so, more precise predictions are possible in the future.[221] The assessment of the prognosis power of the model, however, disproves this claim. By and large, migration potential is turned into what Thomas Krohn (2011) calls a "possibilistic" logic: logic characterized by a distinct vagueness, "fuzziness" and the fact that logical statements refer to a (however small) possibility instead of making a serious attempt at estimating the likelihood of a given event. Like the assumption of a growing migration potential due to the "volcanic outbreak differential" between Germany and Africa, this construction makes possibilistic claims immune to falsification against empirical data, since by definition contradicting evidence can be ignored.[222]

## Conclusion

To summarize, migration potential is one of the few exceptions to the rule that practically applicable knowledge produced by the BAMF cannot make overt reference to academic theory. In contrast to this, migration potential follows a textbook script of the elements of constructing a scientific theory, including definitions, hypothesis, a review of the relevant literature, and cause-event relationships. By itself, this technique is not surprising: the migration potential project was largely self-commissioned, drawn up by scientists with academic training; the theoretical background applied can be considered thoroughly mainstream and not particularly innovative. Barlösius (2008) identifies this as a rather typical feature of governmental knowledge production on the assumption of a risk-avoiding strategy on the side of the ministry:

> "it is rather not necessary, in the contrary even a risk, if departmental research is positioned at the 'peak of science', because its methods and interpretations are often discussed in a controversial manner within academy. The use of such research results threatens a scientific dispute [...] which could disable political action rather than support it. To minimize this risk it is more favorable to the ministry to use secured, undisputable scientific knowledge and according methods, which are part of the established scientific tool box."[223]

221 Bundesamt für Migration und Flüchtlinge 2009b, 23f.

222 Cp. Kron and Winter 2011, p. 211

223 Barlösius 2008, 15f.

By analyzing the political use of the knowledge, the practical application of this theoretical knowledge is quite remarkable: what started out as a largely self-commissioned project to deliver prognoses of future migration movements became a legitimization strategy of migration policy-making. While there are methodological reasons for this scaling back of the research goal, and ultimately at the end, of the practical usability of the concept, the point is that this reduction was counterbalanced with a more elaborate theoretical foundation. It is counter-intuitive to assume a poorer performance in prediction ability from a more elaborate theoretical model rather than the opposite, and yet, this can be assumed in this case. In other words, the theoretical knowledge created in this context is not used to develop the power of analysis any further. The practical usefulness of the concept does not lie in its analytical power, but rather in its remarkable flexibility to legitimize policy.

# The Revenge of Practical Relevance

The analysis of knowledge production at the Federal Office for Migration and Refugees centered on a notion of governmental knowledge: What is it what the state wants to know, how is knowledge produced and what are its core features? As a discussion of relevant theoretical contributions demonstrated, governmental knowledge is somewhat elusive: Governmental knowledge is usually either conceptualized as single pieces of information, or a grand narrative of rationality, discipline or control.[1] In contrast to this, this study focused on carving out the intermediary zones between these two extreme levels of analysis.

If there is a single most important finding of this thesis, then it is the (not overly surprising) fact that the relationship between science and politics has to be carefully scrutinized and evaluated. Specifically, the idea of treating science and politics like two independent spheres seems misleading in the case of the BAMF Research Group and its mission to produce politically relevant knowledge.

What does that mean for the study of governmental knowledge production? In the history of governmental research, some effects and variants of the interconnectedness between politics and science have been described, following broadly four phases of migration research in Germany (Refugee research, "Guest Worker" research, "Lost Decade" and integration and migration research, respectively). In the analysis, empiric evidence was used to draw a differentiated picture which highlights the various interconnections between governance and knowledge production as well as the numerous contradictions, cracks and shifts within and across governmental organizations. As a result, the developments in knowledge production can be linked to according shifts in the governmental logic behind them. Refugee and Ethnic German

1 Cp. Walters 2015, p. 5

migration research and policy-making was characterized by elements of biopolitics to enhance the productive forces of the population, coupled with an according strategy of reporting and knowledge production most closely resembling the instrumentalist approach to knowledge utilization. In contrast to this, "Guest Worker" research and policy-making can be characterized like a technocratic policy complex, which was governed by macro-economic data and according administrative measures. This policy style stands in connection to according principles in the Ministry for Labor and Social affairs, the central coordinating actor in migration policy-making in this era. The competition for influence with the Ministry of the Interior increased especially during the "Lost Decade" in the 1980s and 1990s: After retaining the coordinating role in migration policy-making from the Ministry of Labor and Social Affairs, the Ministry of the Interior developed return-oriented policy principles according to the "no country of immigration" dogma; according knowledge production centered on images of deficit and risk and a strategy of denial of adversary knowledge, the latter particularly arising from a growing academic interest in migration during the 1980s. During that time, two large political camps were established, which each cultivated their own styles of knowledge production: The Ministry of the Interior, large parts of the CDU and CSU conservative parties, on the one hand; the Ministry of Labor and Social affairs, the Commissioner for Foreigners, Worker Unions, churches, the Social-Democratic and Green parties as well as a minority of the conservative CDU on the other.[2] In short, the history of governmental migration research displays a wide array of different political constellations, policy aims and according knowledge production. This confirms the governmentality hypothesis of interconnected processes of knowledge production and political decision-making.

Besides providing historical background as well as examples for different knowledge-power complexes, the history of governmental migration research is important in explaining the so called "paradigm shift". Around the turn of the millennium, legal and administrative reforms which ultimately led to the foundation of the Research Group were triggered by an expert commission on policy reform, the Independent Commission Integration. Its reform proposals seem like a counter-draft to the history of migration research and policy-making, especially the so-called "Lost Decade": Policy-making and research in that era have been portrayed as increasingly antagonistic, where "irrational" policy-makers repeatedly failed to recognize "objective" scientific facts and

2 Gusy and Müller 2012, 4 ff., Herbert 2000, p. 278

act accordingly. In contrast to this, future migration policy-making was to be governed by scientific experts, which decide upon immigration quotas using scientific standards.

These reform proposals were met with considerable resistance, most importantly from the Ministry of the Interior which was keen to retain its central position in migration policy making. As a result, most elements of independent research as well as systematic feedback of knowledge into the political process were removed from the Residence Act. This process is also illustrative for the relationship between knowledge production and policy-making in general: First, the antagonistic picture of "objective science vs. "irrational" politics is misleading, since political claims of any era and of any kind are always founded on arguments and knowledge. What has been called later on the dogma of German migration policy – Germany was not a country of immigration – has been defended against empirical reality by many knowledge producers inside and outside the state bureaucracy. Second, the oft-lamented lack of political influence of scientists is in this example not just a result of systematic differences, bureaucratic sluggishness, or the result of translation costs. Rather, it is the result of an according political strategy, against direct recommendations from a government commission.

Taken together, the history of governmental migration research shows that certain types of knowledge do in fact exercise political influence, while other forms of knowledge are actively and consciously locked out of the political process. For the production of governmental knowledge, this process had some important implications which resulted in a very specific, if not unique, arrangement: First, research was incorporated into a bureaucratic agency, which meant that the researchers were considered for a time as "foreign bodies". This was especially important since the office in question, the BAMF, represented like no other the "Germany is not a Country of Immigration dogma"[3]; its staff and working principles were for at least the first couple of years somewhat overstretched with the new responsibilities resulting from the reform. At the time of foundation of the Research Group, the BAMF was considered an "institutional backwater"[4] which offered little career perspectives for its employees.[5]

---

3 Castles 1985

4 Boswell 2009b, p. 163

5 Field notes, October 2013

Second, despite the principle of hierarchical control, the actual research mandate was unclear from the onset. This was the result of an unspecific legal expression, which led to a process of carving out an area of responsibility for the Research Group: What topics should be analyzed, for what purpose and for whom? As the result of a learning process, the Research Group adopted a quite successful strategy of mimicking departmental research: Research topics are established in a working group involving the ministerial bureaucracy and the BAMF. The *Ausländerzentralregister*, a data base exclusively available to the Research Group, is used extensively as a unique selling point for the BAMF's knowledge production. As a result, staff and financial resources of the Research Group continually expanded, rising to about 25 staff and 400,000 Euros research budget per year in 2013.[6] With this, the BAMF currently plays in the same league like the top tier of migration research institutions in Germany.[7]

Third, as a result of this pragmatic integration into the administration, a specific understanding of governmental research is formulated. Again in the words of a BAMF researcher which have been quoted in the text above:

> "We work flexibly with what serves best. [If] we have a concrete question, we look which methods we can use to answer the question posed to us. In this we are not overly committed to a specific theoretical concept. If we refer to definitions [e.g. in the National Migration Report], these relate to statistical data, and the statistical data depends on legal regulations."[8]

In other words, governmental research is characterized by BAMF researchers as practical (in contrast to theory-oriented abstraction), flexible (instead of methodological rigor) and pragmatic (instead of foundational criticism).

To be clear, at least on the surface, this knowledge is no less "scientific" or "rational" than classic academic knowledge production: Empiric data is collected with scientific methods, analysis and at least in part theory references

6 Email Memo from the Research Group, February 2014

7 Schimany and Schock 2012

8 "Wir arbeiten flexibel mit dem was da ist. [...] Wir haben eine konkrete Frage, wir schauen uns an mit welchen Methoden wir die konkrete Frage die uns gestellt wird beantworten können. Und sind nicht übertrieben eng hinter [...] einem Theoriekonzept her. Wenn wir Definitionen benutzen [zB. im Migrationsbericht] richtet sich bei uns nach den statistischen Erhebungen, die statistischen Erhebungen wiederum richten sich nach dem was in unseren Gesetzen drin steht."(Interview with a BAMF researcher, 2015)

follow established academic principles. Publications are structured and referenced according to academic standards as well; a growing list of academic publications and conference invitations testifies to the fact that the Research Group's knowledge production is acknowledged in the academic world. There is, however, a grave difference between the Research Group and academia when it comes to criteria what counts as good and useful knowledge: the most important quality criterion for governmental knowledge is political relevance. This touches on a theory discussion about the relationship between policy-making and knowledge production. The Research Group's claim of providing politically relevant knowledge refers to a mainstream theory of instrumental knowledge use, which states that research is primarily valued for its informational content. However, in theoretical contributions as well as in the empirical literature, little evidence for instrumental knowledge use is found, which is why this thesis centers on the question what exactly practical relevance signifies, and how it is produced. The hypothesis was that political relevance in the academic literature is usually conceptualized as either a direct and measurable influence of research on political decisions, or as structural features of the research-policy system that ensure the systematic feedback of expert knowledge on political decisions.[9] Since this does not mirror the understanding of policy relevance by the involved actors, a practice-oriented understanding of political relevance has been developed. In chapter 3.3, the institutional process has been described which is characterized by a constant learning process on the side of the researchers and a long-term strategy of acquiring study commissions from state actors. Based on this, an alternative understanding of political relevance was developed which gives credit to the fact that the Research Group has successfully adapted its research output to demand by other state actors.

In this understanding, practical relevance is not an abstract quality criterion, but depends on the concrete political practice of the respective contractor, the research topic, and current political measures in the field. In the analysis, several practices have been outlined, such as the provision of legibility, depoliticization, calming of the public debate, and legitimization. In this sense, not only the direct influence of a political decision is analyzed, but also strategies which might have failed, or might have been altered in the meantime in answer to political changes in the field. For example, political relevance in integration research is subject to shifts in the governmental logic behind

9 Cp. Scholten et al. 2015a

the knowledge: While initial demand for integration research resulted in an Esser-inspired theory of assimilation and deficit-orientation, this approach was altered in the course of increased legislative activities to recruit qualified workers from third countries since 2012. Integration of these migrants is oriented towards enabling migration termed as "Welcome Culture".[10]

In sum, despite the external limitations, the Research Group has systematically created and defended an area of competence and influence both vis-a-vis peer knowledge producers and the political-administrative system and does in fact provide politically relevant knowledge. In this context, the widespread hypothesis of a systematic gap between research and bureaucracy has to be reevaluated: The initial isolation of the Research Group, expressed in the fact that research tasks were misunderstood, as well as a feeling of estrangement vis-a-vis government officials has successfully been overcome. This was not caused by systematic differences but by the specific situation of institutional change and a rather blurry legal mandate of the Research Group.[11]

This pragmatic understanding of the production of politically relevant knowledge comes however at a cost. The general strategy of integration into the state bureaucracy signifies on the one hand the agency of researchers which successfully navigate in the administrative structure of the BAMF and seek strategic opportunities for the provision of politically relevant knowledge. On the other hand, practical relevance makes research vulnerable to political manipulation, since politically relevant research questions are often formulated in a partisan way to support specific ex-ante policy preferences. This vulnerability is augmented by the rather precarious institutional status of the Research Group as an in-house unit of an administrative authority: research agenda setting and publication are subject to hierarchical supervision, so that the Research Group depends on the good will of the Ministry of the Interior as one interlocutor remarked;[12] this is at the same time the key difference to other departmental research institutes, which enjoy more institutional independence. This means that usually, research is conducted within narrowly defined borders which cannot be questioned: Integration, for example, is understood as the participation of migrants in various integration policy instruments, not as an onmi-societal process of transformation. In gen-

10 Bundesamt für Migration und Flüchtlinge 2013d, Heckmann 2012

11 Kraler and Perchinig 2017, p. 85

12 Interview with a BAMF researcher, 2016

eral, political and legal concepts are being operationalized, differentiated and diversified by the BAMF research; they are however never revised as such or evaluated critically. This becomes most clearly apparent in the defensive strategy of avoiding negative feedback, which is tactically employed in politically heated research topics, such as Muslims, naturalization, and the like. Quite ironically, as a consequence, the idealistic role of experts as portrayed by the Independent Commission Immigration is the weakest exactly in cases where scientific recommendations are needed most to overcome ideologically fueled political deadlock. By and large, the knowledge produced at the BAMF buys its political relevance through uncontroversiality, affirmation and discursive legitimization of political and administration decisions.[13]

This "revenge of practical relevance" is of course not unique to the Research Group since all knowledge producers face the same basic dilemma: How to produce knowledge which is both politically relevant and objective?[14] The four knowledge-power complexes discussed here offer distinct case examples in this regard, all of which answer this question differently: Administrative knowledge is reduced to statistical reporting and legislative definitions, which are used as a basis for the establishment of Migrant Groups as a new statistical concept. With the increasing focus on selected target groups, the Migration Reports form a specific perspective of governmentality, supporting the image that migration is an orderly social process under the control of the government. Knowledge on Muslims and other politically controversial topics is geared towards calming public debate by retreating to apolitcal, technical positions and therefore produce a standard of objectivity.[15] In the example of integration research, initial knowledge production can be characterized rather theory-driven by implementing a hegemonic approach adopted from Hartmut Esser's assimilation theory. Migration potential is developed from a prognosis instrument to a self-referential legitimization strategy. Again, all these knowledge-power complexes demonstrate that the Research Group did in fact deliver politically relevant knowledge: This is especially true for some widely disseminated studies such as the Migration Reports, Muslim Life in Germany or the Integration Panel. Theoretical concepts, such as the assimilationist approach to integration or migration potential have gained widespread acceptance also thanks to the BAMF's research work. This success in political

13 Cp. Hetfleisch 2013

14 Cp. Amir-Moazami 2018b, p. 111. See also Boswell and D'Amato 2012, p. 16

15 Amir-Moazami 2018b, p. 104

relevance, however, is paid for with institutional dependency: Irrespective of the success or failure of the individual research strategies, the analysis of governmental knowledge made the fact clearly visible that none of the knowledge under scrutiny here can be called objective by any standard. While this finding is not overly surprising in the face of the above-mentioned dilemma, it does however stand at odds with the self-proclaimed image of provision of neutral expert knowledge. This follows by and large a defensive rationale of calming the general public: The higher the degree of politicization of a given topic, the more governmental research retreats to a technical, apolitical point of view. This in turn offers additional possibilities for politicization since research results can be used to support *any* political claim.[16]

To deconstruct the claim of neutrality, the context-specific governmental perspective has been described both in contemporary BAMF research fields and in historic migration research. This perspective is shaped by institutional constellations, competition between state actors, material restraints and theoretical ideas which are specific for each of the knowledge-power complexes analyzed here. Connected to this, the claim of political usefulness can be likewise deconstructed if extrapolating from a given study its perceived political applicability. Concerning the latter, the analysis revealed that political usefulness cannot be regarded a uniform feature of knowledge, or a yardstick of epistemic quality, as sometimes suggested.[17] Rather, there are different potential political uses to which the BAMF's knowledge can be applied; some of which have been described here.

While the analysis has demonstrated how the individual strategies have been carved out, and have sometimes been successful in reaching their aims, it seems clear that policy relevance also puts a strain on the epistemic quality of the knowledge produced. In the case of the migration potential as well as integration research, the requirement of political relevance can be connected to specific bias sources in the knowledge structure: The relatively one-sided analysis through a neoclassic push-pull framework renders a coherent legitimization for current EU migration policy. However, it also impedes the systematic generation and testing of hypotheses about future migration movements and therefore, a systematic approach to enhance the quality of the knowledge in the long term. In fact, by standards of systematic analysis,

16 Heckmann and Wiest 2015, 198f.

17 Mayr et al. 2011

the prediction potential of the analysis framework of migration potential has decreased over time.[18]

Similarly, in the field of integration research, the construction of a coherent integration theory contributed to a depoliticizing of the formerly most contested policy field in the area, and secured valuable areas of competence for the BAMF. However, this success came at the cost of representing integration as unilateral individual effort of the migrant which systematically disregards structural barriers to access to social and economic resources. At the same time, the more recent differentiated analysis frameworks for privileged migrant groups introduce incoherencies into the hitherto uniform knowledge order of integration. By selectively applying this model to "problematic" immigrant groups, such as Muslims, Integration Course participants, or immigrated spouses, knowledge production shapes a particular image of integration which disregards structural factors of exclusion and places the responsibility for integration solely at the hands of the immigrants. In fact, in the first years of the Research Group's existence, there have been no publications dedicated to discrimination or racism, neither in-house nor externally. This complete neglect can be regarded at the same time the largest difference as compared to academic integration research, in which discrimination and racism feature among the most important research topics in migration research.[19] By contrast to this, immigrant groups which are perceived useful such as university graduates, high-skilled or self-employed migrants, structural barriers to integration (for example, excess bureaucracy, discrimination, etc.) are part of the framework of analysis.

Another useful example here is the governmentality discourse created in the Migration Reports: the politically useful image of migration as a steered, orderly process can only be created if the single most important migration form is excluded from analysis. Also, in this case, epistemic quality is sacrificed for a less contradictory, more coherent and thus more politically useful narrative. In the context of migration potential, a similar selective application of theory can be discerned: While studies on the potential of migration from Africa and Eastern Europe conclude that migration is a harmful, uncontrollable danger, the same processes are evaluated quite positively in the context of intra-EU migration. This different conclusion is based to a large degree on a selective application of theory and according data which confirm the ex-

18 Tetlock 2005, 47ff.

19 Leibnitz-Institut für Sozialwissenschaften 2010, 12f.

ante assumption. Here, knowledge is useful since it supports the general EU migration strategy: While intra-EU migration is supported, migration from outside the EU is considered harmful and has to be curbed as much as possible. The respective migration potential studies deliver a well-suited scientific foundation for this arrangement which make the political choice seem like a scientifically grounded decision.

All of these problematic developments can be connected to one common root cause: the lack of proper theoretical discussion and theory development. As already mentioned, theory discussion is regarded as art for art's sake in governmental research; the lack of theory development and pragmatic selection of useful concepts is regarded as one core pillar to the provision of practical relevance, as stated by the BAMF.[20] As a result, governmental knowledge perpetuates uncontroversial mainstream theory (which can be outdated), reads theory too narrowly (as in the case of integration concepts), and develops blind spots and taboos (as in the case of discrimination). The result is common-sensical knowledge which reveals its inherent inconsistencies if new immigrant groups challenge to uniform picture, such as the different integration paradigm for economically attractive migrants versus those which are considered problematic, or the different migration potential discussion of African and intra-European migration.

All in all, the Research Group contributes with its knowledge to an image of the state as a keeper of the common good: The state keeps an overview, it demands integration and provides support to it, it provides objective information for heated political topics and it protects the borders from threats.

20 Bundesamt für Migration und Flüchtlinge 2015a, p. 22

# Appendix

## Bibliography


Affolter, Laura (2017): "Asyl-Verwaltung kraft Wissen. Die Herstellung von Entscheidungswissen in einer Schweizer Asylbehörde," In Christian Lahusen, Stephanie Schneider (Eds.), Asyl verwalten. Zur bürokratischen Bearbeitung eines gesellschaftlichen Problems. Bielefeld: transcript (Kultur und soziale Praxis), pp. 145–173.

AG Ressortforschungseinrichtungen (2016): "Mitglieder der AG Ressortforschungseinrichtungen", updated on 9/26/2016, checked on 2/13/2017.

Amir-Moazami, Schirin (2018a): "Der inspizierte Muslim. Zur Politisierung der Islamforschung in Europa", Bielefeld: transcript.

Amir-Moazami, Schirin (2018b): "Epistemologien der 'muslimischen Frage' in Europa", In Schirin Amir-Moazami (Ed.), Der inspizierte Muslim. Zur Politisierung der Islamforschung in Europa, Bielefeld: transcript, pp. 91–123.

Angenendt, Steffen (1992): "Ausländerforschung in Frankreich und der Bundesrepublik Deutschland. Gesellschaftliche Rahmenbedingungen und inhaltliche Entwicklung eines aktuellen Forschungsbereichs", Frankfurt, New York: campus.

Angenendt, Steffen (2012): "Migration, Mobilität und Entwicklung. EU-Mobilitätspartnerschaften als Instrument der Entwicklungszusammenarbeit". Berlin.

Aumüller, Jutta (2009): "Assimilation. Kontroversen um ein migrationspolitisches Konzept", Bielefeld: transcript.

Bade, Klaus (Ed.) (1994): "Das Manifest der 60. Deutschland und die Einwanderung", München: C. H. Beck.

Bade, Klaus (2001): "Konzeptionsentwurf zur Institutionellen Strukturierung des Migrationswesens unter besonderer Berücksichtigung der Organisation der Migrationsforschung in Deutschland. Gutachten für die Unabhängige Kommission Zuwanderung", Berlin.

Bade, Klaus (2013): "Zwanzig Jahre 'Manifest der 60. Deutschland und die Einwanderung' (1993). Erinnerungen des Herausgebers", in: Rat für Migration, Jüdisches Museum Berlin (Eds.), "Dokumentation der Tagung Migrations- und Integrationspolitik heute", Berlin.

Bade, Klaus (2016): "Von Unworten zu Untaten. Kulturängste, Populismus und politische Feindbilder in der deutschen Migrations- und Asyldiskussion zwischen 'Gastarbeiterfrage' und 'Flüchtlingskrise'" in: *IMIS Beiträge* 48, pp. 35–171.

Bade, Klaus (2017): "Migration – Flucht – Integration. Kritische Politikbegleitung von der 'Gastarbeiterfrage' bis zur 'Flüchtlingskrise'. Erinnerungen und Beiträge," Karlsruhe: Loeper.

Barlösius, Eva (2008): "Zwischen Wissenschaft und Staat? Die Verortung der Ressortforschung", in: *Wissenschaftszentrum Berlin Discussion papers* 101.

Bax, Daniel (2012): "Die Betroffenen hätten lieber beide. Umfrage zur Pass-Entscheidungspflicht", in: *Tageszeitung*, 6/22/2012.

Beauftragte der Bundesregierung für Ausländerfragen (1997): "Bericht der Beauftragten der Bundesregierung für Ausländerfragen über die Lage der Ausländer in der Bundesrepublik Deutschland". (=Ausländerbericht, 3), Bonn.

Beauftragte der Bundesregierung für Migration, Flüchtlinge und Integration (2005): "Bericht der Beauftragten der Bundesregierung für Migration, Flüchtlinge und Integration über die Lage der Ausländerinnen und Ausländer in Deutschland. Beauftragte der Bundesregierung für Migration, Flüchtlinge und Integration, (=Ausländerbericht, 6), Berlin.

Beck, Ulrich (2000): "Freiheit oder Kapitalismus. Gesellschaft neu denken", Frankfurt am Main: Suhrkamp.

Beck, Ulrich (2004): "Der kosmopolitische Blick. oder: Krieg ist Frieden", Frankfurt: Suhrkamp.

Beer, Mathias (2003): "Symbolische Politik? Entstehung, Aufgaben und Funktion des Bundesministeriums für Vertriebene, Flüchtlinge und Kriegsgeschädigte", in: Jochen Oltmer (Ed.), "Migration steuern und verwalten. Deutschland vom späten 19. Jahrhundert bis zur Gegenwart", Osnabrück: V&R unipress, pp. 295–322.

Bil, Rob; Verweij, Arjen (Eds.) (2012): "Measuring and monitoring immigrant integration in Europe. The Nederlands Institute for Social Research", Den Haag.

Bommes, Michael (2009): "Migration Reseach in Germany: The Emergence of a Generalised Field in a Reluctant Immigrant Country", in: Dietrich Thränhardt, Michael Bommes (Eds.), "National Paradigms of Migration Research", Osnabrück: V&R unipress, pp. 127–186.

Bommes, Michael (2012): "Migration in modern society", in: Christina Boswell, Gianni D'Amato (Eds.), "Immigration and Social Systems", Amsterdam: Amsterdam University Press, pp. 19–36.

Bommes, Michael; Thränhardt, Dietrich (2012): "National paradigms of migration research", In Christina Boswell, Gianni D'Amato (Eds.), "Immigration and Social Systems", Amsterdam: Amsterdam University Press, pp. 201–232.

Boswell, Christina (2008): "The political functions of expert knowledge: knowledge and legitimation in European Union immigration policy", in: *Journal of European Public Policy* 15, pp. 471–488.

Boswell, Christina (2009a): "Knowledge, Legitimation and the Politics of Risk. The Functions of Research in Public Debates on Migration", in: *Political Studies* 57, pp. 165–186.

Boswell, Christina (2009b): "The Political Uses of Expert Knowledge. Immigration Policy and Social Research", Cambridge: Cambridge University Press.

Boswell, Christina (2011): "Migration Control and Narratives of Steering", in: *The British Journal of Politics & International Relations* 13, pp. 12–25.

Boswell, Christina (2015): "Cultures of Knowledge Use in Policymaking: The Functions of Research in German and UK Immigration Policy", in: Peter Scholten, Han Entzinger, Rinus Penninx, Stijn Verbeek (Eds.), "Integrating Immigrants in Europe", Heidelberg: Springer, pp. 19–38.

Boswell, Christina (2018): "Manufacturing political trust: targets and performance management in public policy", Cambridge: Cambridge University Press.

Boswell, Christina; D'Amato, Gianni (2012): "Introduction", in: Christina Boswell, Gianni D'Amato (Eds.), "Immigration and Social Systems", Amsterdam: Amsterdam University Press, pp. 11–18.

Boswell, Christina; Geddes, Andrew; Scholten, Peter (2011): "The Role of Narratives in Migration Policy-Making. A Research Framework", in: *The British Journal of Politics & International Relations* 13, pp. 1–11.

Bourdieu, Pierre (1977): "Outline of a Theory of Practice", Cambridge: Cambridge University Press.

Brücker, Herbert; Bass, Timo (2010): "Wirkungen der Zuwanderungen aus den neuen mittel- und osteuropäischen EU-Staaten auf Arbeitsmarkt und Gesamtwirtschaft", in: *WISO Diskurs*.

Bubenhofer, Noah: Einführung in die Korpuslinguistik. Available online at http://www.bubenhofer.com/korpuslinguistik/kurs/, checked on 12/11/2015.

Bundesamt für Migration und Flüchtlinge (Ed.) (2005a): "Der Einfluss von Zuwanderung auf die deutsche Gesellschaft. Deutscher Beitrag zur Pilotforschungsstudie "The Impact of Immigration on Europe's Societies" im Rahmen des Europäischen Migrationsnetzwerks (= BAMF Forschungsbericht, 1), Nürnberg.

Bundesamt für Migration und Flüchtlinge (2005b): "Illegalität von Migranten in Deutschland. Zusammenfassung des Forschungsstandes." (= BAMF Working Paper No. 2), Nürnberg.

Bundesamt für Migration und Flüchtlinge (2005c): "Jüdische Zuwanderer in Deutschland. Ein Überblick über den Stand der Forschung." (= BAMF Working Paper No. 3), Nürnberg.

Bundesamt für Migration und Flüchtlinge (2006): "Migrationsbericht 2005. Des Bundesamtes für Migration und Flüchtlinge im Auftrag der Bundesregierung", Nürnberg.

Bundesamt für Migration und Flüchtlinge (2007a): "Migrationsbericht 2006 des Bundesamtes für Migration und Flüchtlinge. im Auftrag der Bundesregierung", Nürnberg.

Bundesamt für Migration und Flüchtlinge (2007b): "Soziodemographische Merkmale, Berufsstruktur und Verwandtschaftsnetzwerke jüdischer Zuwanderer", (= BAMF Working Paper No. 8), Nürnberg.

Bundesamt für Migration und Flüchtlinge (2007c): "Zuwanderung und Integration von (Spät-)Aussiedlern. Ermittlung und Bewertung der Auswirkungen des Wohnortzuweisungsgesetzes" (=BAMF Forschungsbericht, 3), Nürnberg.

Bundesamt für Migration und Flüchtlinge (2008a): "Das Integrationspanel. Ergebnisse zur Integration von Teilnehmern zu Beginn ihres Integrationskurses. (= BAMF Working Paper No. 19), Nürnberg.

Bundesamt für Migration und Flüchtlinge (2008b): "Die Einbürgerung von Ausländern in Deutschland. Aus der Reihe "Integrationsreport", Teil 3", (= BAMF Working Paper No. 17), Nürnberg.

Bundesamt für Migration und Flüchtlinge (2008c): "Healthy-Migrant-Effect, Erfassungsfehler und andere Schwierigkeiten bei der Analyse der Morta-

lität von Migranten. Eine Bestandsaufnahme", (= BAMF Working Paper No. 15), Nürnberg.
Bundesamt für Migration und Flüchtlinge (Ed.) (2008d): "Leitbild BAMF."
Bundesamt für Migration und Flüchtlinge (2008e): "Schulische Bildung von Migranten in Deutschland. Aus der Reihe 'Integrationsreport' Teil 1", (= BAMF Working Paper No. 13), Nürnberg.
Bundesamt für Migration und Flüchtlinge (2008f): "Sprachliche Integration von Migranten in Deutschland. Aus der Reihe ,Integrationsreport', Teil 2", (= BAMF Working Paper No. 14), Nürnberg.
Bundesamt für Migration und Flüchtlinge (2008g): "Türkische, griechische, italienische und polnische Personen sowie Personen aus den Nachfolgestaaten des ehemaligen Jugoslawien in Deutschland. Erste Ergebnisse der Repräsentativbefragung 'Ausgewählte Migrantengruppen in Deutschland 2006/2007' (RAM)," (= BAMF Working Paper No. 11), Nürnberg.
Bundesamt für Migration und Flüchtlinge (2008h): "Wohnen und innerstädtische Segregation von Migranten in Deutschland. Aus der Reihe 'Integrationsreport' Teil 4." (= BAMF Working Paper No. 21), Nürnberg.
Bundesamt für Migration und Flüchtlinge (2009a): "Das Integrationspanel. Entwicklung von alltagsrelevanten Sprachfertigkeiten und Sprachkompetenzen der Integrationskursteilnehmer während des Kurses," (= BAMF Working Paper No. 23), Nürnberg.
Bundesamt für Migration und Flüchtlinge (2009b): "Migrations- und Integrationsforschung. Tätigkeitsbericht 2007/2008 der Forschungsgruppe im Bundesamt für Migration und Flüchtlinge," Nürnberg.
Bundesamt für Migration und Flüchtlinge (2009c): "Muslimisches Leben in Deutschland. im Auftrag der Deutschen Islam Konferenz," (= BAMF Forschungsbericht, 6), Nürnberg.
Bundesamt für Migration und Flüchtlinge (2009d): "Vor den Toren Europas? Das Potenzial der Migration aus Afrika," (= BAMF Forschungsbericht, 7), Nürnberg.
Bundesamt für Migration und Flüchtlinge (2009e): "Zuwanderung von Hochqualifizierten aus Drittstaaten nach Deutschland. Ergebnisse einer schriftlichen Befragung. " (= BAMF Working Paper No. 28), Nürnberg.
Bundesamt für Migration und Flüchtlinge (2010a): "Mediennutzung von Migranten in Deutschland. Integrationsreport 8", (= BAMF Working Paper No. 34), Nürnberg.

Bundesamt für Migration und Flüchtlinge (2010b): "Migrations- und Integrationsforschung. Jahresbericht 2009 der Forschungsgruppe im BAMF", Nürnberg.

Bundesamt für Migration und Flüchtlinge (2010c): "Repräsentativbefragung "Ausgewählte Migrantengruppen in Deutschland 2006/2007", Nürnberg.

Bundesamt für Migration und Flüchtlinge (2011a): "Das Integrationspanel. Ergebnisse einer Längsschnittstudie zur Wirksamkeit und Nachhaltigkeit von Integrationskursen", (= BAMF Forschungsbericht, 11), Nürnberg.

Bundesamt für Migration und Flüchtlinge (2011b): "Der Einfluss des Integrationskurses auf die Integration russisch- und türkischstämmiger Integrationskursteilnehmerinnen. Qualitative Ergänzungsstudie zum Integrationspanel", (= BAMF Working Paper No. 37), Nürnberg.

Bundesamt für Migration und Flüchtlinge (2011c): "Jahresbericht 2010 der Forschungsgruppe im Bundesamt für Migration und Flüchtlinge", Nürnberg.

Bundesamt für Migration und Flüchtlinge (2011d): "Migranten am Arbeitsmarkt in Deutschland. Integrationsreport 9", (= BAMF Working Paper No. 36), Nürnberg.

Bundesamt für Migration und Flüchtlinge (2011e): "Migranten im Niedriglohnsektor unter besonderer Berücksichtigung der Geduldeten und Bleibeberechtigten". (= BAMF Working Paper No. 39), Nürnberg.

Bundesamt für Migration und Flüchtlinge (2011f): "Migrationsbericht 2010", Nürnberg.

Bundesamt für Migration und Flüchtlinge (2011g): "Potenziale der Migration zwischen Afrika und Deutschland", (= Beiträge zu Migration und Integration, 2), Nürnberg.

Bundesamt für Migration und Flüchtlinge (2012a): "Das Migrationspotenzial aus der GUS in die Europäische Union," (BAMF Forschungsbericht, 17), Nürnberg.

Bundesamt für Migration und Flüchtlinge (2012b): "Die Optionsregelung im Staatsangehörigkeitsrecht aus der Sicht von Betroffenen," (BAMF Forschungsbericht, 16), Nürnberg.

Bundesamt für Migration und Flüchtlinge (2012c): "Die Organisation der Asyl- und Zuwanderungspolitik in Deutschland. Studie der deutschen Kontaktstelle für das Europäische Migrationsnetzwerk (EMN)", (= BAMF Working Paper No. 25), Nürnberg.

Bundesamt für Migration und Flüchtlinge (2012d): "Einbürgerungsverhalten von Ausländerinnen und Ausländern in Deutschland sowie Erkenntnisse zu Optionspflichtigen. Ergebnisse der BAMF-Einbürgerungsstudie 2011," (BAMF Forschungsbericht, 15), Nürnberg.

Bundesamt für Migration und Flüchtlinge (2012e): "Islamisches Gemeindeleben in Deutschland. im Auftrag der Deutschen Islam Konferenz," (BAMF Forschungsbericht, 13), Nürnberg.

Bundesamt für Migration und Flüchtlinge (2012f): "Migrations- und Integrationsforschung. Jahresbericht 2011 der Forschungsgruppe im Bundesamt für Migration und Flüchtlinge", Nürnberg.

Bundesamt für Migration und Flüchtlinge (2012g): "Politische Einstellungen und politische Partizipation von Migranten in Deutschland. aus der Reihe "Integrationsreport", Teil 10", (= BAMF Working Paper No. 46), Nürnberg.

Bundesamt für Migration und Flüchtlinge (2012h): "Zuwanderung von internationalen Studierenden aus Drittstaaten. Studie der deutschen nationalen Kontaktstelle für das Europäische Migrationsnetzwerk (EMN) ", (= BAMF Working Paper No. 47), Nürnberg.

Bundesamt für Migration und Flüchtlinge (2012i): "Zuwanderung von selbständigen und freiberuflichen Migranten aus Drittstaaten nach Deutschland. Ergebnisse einer schriftlichen Befragung von Selbständigen und Freiberuflern nach § 21 AufenthG", (= BAMF Working Paper No. 48), Nürnberg.

Bundesamt für Migration und Flüchtlinge (Ed.) (2013a): „60 Jahre Bundesamt für Migration und Flüchtlinge: Festschrift Forschung", Nürnberg.

Bundesamt für Migration und Flüchtlinge (2013b): "Das Integrationspanel. Langfristige Integrationsverläufe von ehemaligen Teilnehmenden an Integrationskursen", (= BAMF Working Paper No. 52), Nürnberg.

Bundesamt für Migration und Flüchtlinge (2013c): „(Spät-)Aussiedler in Deutschland. Eine Analyse aktueller Daten und Forschungsergebnisse (= BAMF Forschungsbericht, 20), Nürnberg.

Bundesamt für Migration und Flüchtlinge (2013d): "Willkommens- und Anerkennungskultur. Handlungsempfehlungen und Praxisbeispiele", Nürnberg.

Bundesamt für Migration und Flüchtlinge (2014a): "Beschäftigung ausländischer Absolventen deutscher Hochschulen. Ergebnisse der BAMF-Absolventenstudie 2013" (= BAMF Forschungsbericht, 23), Nürnberg.

Bundesamt für Migration und Flüchtlinge (2014b): "Die Integration von zugewanderten Ehegattinnen und Ehegatten in Deutschland. BAMF-Heiratsmigrationsstudie 2013" (= BAMF Forschungsbericht, 22), Nürnberg.

Bundesamt für Migration und Flüchtlinge (2014c): "Geschlechterrollen bei Deutschen und Zuwanderern christlicher und muslimischer Religionszugehörigkeit" (= BAMF Forschungsbericht, 21), Nürnberg.

Bundesamt für Migration und Flüchtlinge (2014d): "Migrations- und Integrationsforschung. Jahresbericht 2013 des Forschungszentrums Migration, Integration und Asyl im Bundesamt für Migration und Flüchtlinge", Nürnberg.

Bundesamt für Migration und Flüchtlinge (2014e): "Zuwanderung aus den neuen EU-Mitgliedstaaten Bulgarien und Rumänien" (= BAMF Forschungsbericht, 24), Nürnberg.

Bundesamt für Migration und Flüchtlinge (2015a): „10 Jahre Integrationsarbeit des Bundesamtes für Migration und Flüchtlinge", Nürnberg

Bundesamt für Migration und Flüchtlinge (2015b): "Das Potenzial der Migration aus Indien. Entwicklungen im Herkunftsland, internationale Migrationsbewegungen und Migration nach Deutschland" (= Forschungsbericht, 26), Nürnberg.

Bundesamt für Migration und Flüchtlinge (2015c): "Forschungszentrum für Migration, Integration und Asyl, Gesamtliste der Publikationen", Nürnberg.

Bundesamt für Migration und Flüchtlinge (2015d): „10 Jahre BAMF Forschungszentrum. Availible online at https://www.bamf.de/DE/DasBAMF/10JahreIntegrationsarbeit/10JahreForschung/10Jahre_forschung-node.html. Last checked on 18.1.2018

Bundesamt für Migration und Flüchtlinge (2016a): "IAB-BAMF-SOEP-Befragung von Geflüchteten: Überblick und erste Ergebnisse", Nürnberg.

Bundesamt für Migration und Flüchtlinge (2016b): "Migrations- und Integrationsforschung. Jahresbericht 2015 des Forschungszentrums Migration, Integration und Asyl im Bundesamt für Migration und Flüchtlinge", Nürnberg.

Bundesamt für Migration und Flüchtlinge (2016c): "Wie viele Muslime leben in Deutschland? Eine Hochrechnung über die Anzahl der Muslime in Deutschland zum Stand 31. Dezember 2015", (= BAMF Working Paper No. 71), Nürnberg.

Bundesamt für Migration und Flüchtlinge; Deutsche Islam Konferenz (2017): "Methodenbericht: Muslimisches Leben in Deutschland 2016", Nürnberg.

Bundesanstalt für Arbeit (1964): "Anwerbung und Vermittlung Ausländischer Arbeitnehmer. Erfahrungsbericht 1963", Nürnberg.

Bundesanstalt für Arbeit (1965): "Anwerbung und Vermittlung Ausländischer Arbeitnehmer. Erfahrungsbericht 1964", Nürnberg.
Bundesanstalt für Arbeit (1971): "Ausländische Arbeitnehmer 1970. Beschäftigung, Anwerbung, Vermittlung – Erfahrungsbericht 1970", Nürnberg.
Bundesanstalt für Arbeit (1972): "Ausländische Arbeitnehmer 1971. Beschäftigung, Anwerbung, Vermittlung – Erfahrungsbericht 1971", Nürnberg.
Bundesanstalt für Arbeit (1973): "Repräsentativuntersuchung '72", Nürnberg.
Bundesanstalt für Arbeit (1974): "Ausländische Arbeitnehmer 1972/1973. Beschäftigung, Anwerbung, Vermittlung – Erfahrungsbericht 1972/73", Nürnberg.
Bundesanstalt für Arbeitsvermittlung und Arbeitslosenversicherung (1962): "Anwerbung und Vermittlung Ausländischer Arbeitnehmer, Erfahrungsbericht 1961", Nürnberg.
Der Bundesminister für Arbeit und Sozialordnung (1981): "Situation der ausländischen Arbeitnehmer und ihrer Familienangehörigen in der Bundesrepublik Deutschland" (= Repräsentativuntersuchung '80), Bonn.
Bundesministerium des Innern (1989): "Aufzeichnung zur Ausländerpolitik und zum Ausländerrecht in der Bundesrepublik Deutschland" Bonn.
Bundesministerium des Innern (1/15/2014): "Migrationsbericht 2012. Das Bundeskabinett hat heute den Migrationsbericht 2012 beschlossen und zur weiteren Beratung an den Deutschen Bundestag weitergeleitet", Press release, Berlin.
Bundesministerium des Innern (12/4/2015): "Migrationsbericht 2015. Bundeskabinett verabschiedet Migrationsbericht 2015 und Hochrechnung zur Zahl der Muslime in Deutschland." Press release, Berlin.
Bundesministerium des Innern (2016): "Act on the Residence, Economic Activity and Integration of Foreigners in the Federal Territory", Residence Act, revised 3/11/2016. Source: Bundesgesetzblatt I, p. 394.
Bundesministerium des Innern (1/6/2016): "'Entscheidend ist, wie schnell wir ohne Konflikte zusammenwachsen.' Bundesinnenminister stellt den Migrationsbericht 2014 vor." Press release, Berlin.
Bundesministerium des Innern; Bundesamt für Migration und Flüchtlinge (2014): „Migrationsbericht 2013".
Bundesministerium des Innern; Bundesamt für Migration und Flüchtlinge (2016): "Migrationsbericht 2015 des Bundesamtes für Migration und Flüchtlinge im Auftrag der Bundesregierung", Nürnberg.

Bundesministerium für Arbeit und Sozialordnung (1996): "Situation der ausländischen Arbeitnehmer und ihrer Familienangehörigen in der Bundesrepublik Deutschland", Bonn.

Bundesministerium für Arbeit und Sozialordnung (2001): "Teil A: Türkische, ehemalige jugoslawische, italienische sowie griechische Arbeitnehmer und ihre Familienangehörigen in den alten Bundesländern und im ehemaligen West-Berlin", Berlin.

Bundesministerium für Vertriebene, Flüchtlinge und Kriegsgeschädigte (5/22/1953): "Gesetz über die Angelegenheiten der Vertriebenen und Flüchtlinge. BVFG, revised 1. " Source: Bundesgesetzblatt I, 22.5.1953.

Bundesregierung (2007): "Zehn Leitlinien einer modernen Ressortforschung", Berlin/Bonn

Castles, Stephen (1985): "The Guest Who Stayed. The Debate on 'Foreigners Policy' in the German Federal Republic", in: *International Migration Review* 19, pp. 517–534.

Castles, Stephen (2000): "Ethnicity and Globalization. From Migrant Worker to Transnational Citizen." London: Sage.

Castles, Stephen (2004): "The Factors That Make and Unmake Migration Policies", in: *International Migration Review* 38, pp. 852–884.

Castles, Stephen; Wihtol de Wenden, Catherine (2006): "Framing International Migration: from National Models to Transnational Critique", in: Ellie Vasta, Vasoodeven Vuddamalay (Eds.), "International Migration and the Social Sciences. Confronting National Experiences in Australia, France and Germany", Basingstoke: Palgrave Macmillan, pp. 222–251.

Diehl, Claudia; Fick, Patrick (2012): "Deutschsein auf Probe: Der Umgang deutsch-türkischer junger Erwachsener mit dem Optionsmodell", in: *Soziale Welt* 63, pp. 339–360.

Directorate-General Justice, Freedom and Security (2010): "Handbook on Integration for policy-makers and practitioners", Brussels.

Dohse, Knuth (1981): "Ausländische Arbeiter und bürgerlicher Staat. Genese und Funktion von staatlicher Ausländerpolitik und Ausländerrecht." Königstein/Taunus: Anton Hain.

Doughan, Sultan; Tzuberi, Hannah (2018): "Säkularismus als Praxis und Herrschaft: Zur Kategorisierung von Juden und Muslimen im Kontext säkularer Wissensproduktion", in: Schirin Amir-Moazami (Ed.), "Der inspizierte Muslim. Zur Politisierung der Islamforschung in Europa", Bielefeld: transcript, pp. 269–308.

Eichenhofer, Johannes (2013): "Begriff und Konzept der Integration im Aufenthaltsgesetz", Baden-Baden: Nomos.

Engler, Fabian (2014): "Vertagte Anerkennung. Teilwerdung des Islams und die Grenzen der Zugehörigkeit im politischen Dialog der Deutschen Islam Konferenz", in: Jens Adam, Asta Vonderau (Eds.), "Formationen des Politischen", Bielefeld: transcript, 67-93.

Engler, Fabian (2020): "Islam und Deutschland. Das strittige Gemeinsame im politischen Dialog der Deutschen Islam Konferenz", Europa-Universität Viadrina: Frankfurt/Oder.

Esser, Hartmut (2001): "Integration und ethnische Schichtung", in: *Arbeitspapiere – Mannheimer Zentrum für Europäische Sozialforschung* (40).

Esser, Hartmut (2009): "Wertekonsens und die Integration offener Gesellschaften", in: Deutsche Islam Konferenz (Ed.), "Drei Jahre Deutsche Islam Konferenz (DIK) 2006-2009. Muslime in Deutschland – Deutsche Muslime", Berlin, pp. 82–106.

European Commission (2011): "The Global Approach to Migration and Mobility", Brussels.

European Commission (2014): "Report on the implementation of the Global Approach to Migration and Mobility. 2012-2013", Brussels

Favell, Adrian (2001): "Integration Policy and Integration Research in Europe: A Review and a Critique", in: Thomas Alexander Aleinikoff (Ed.), "Citizenship Today: Global perspectives and practices", Washington: Brookings Institution Press, pp. 349–400.

Ferguson, James (1994): "The Anti-Politics Machine. 'Development', Depolitization, and Bureaucratic Power in Lesotho", Minneapolis: University of Minnesota Press.

Foucault, Michel (2014): "On the Government of the Living. Lectures at the Collège de France, 1979-1980", Basingstoke: Palgrave Macmillan.

Friedrich, Lena; Waibel, Stine (2012): "Local Integration Concepts in Germany – Diffusion of an Integration Model", in: *IMIS Beiträge* (41), pp. 53–72.

Geiger, Martin (2013): "The Transformation of Migration Politics: From Migration Control to Disciplining Mobility", in: Martin Geiger, Antoine Pécoud (Eds.), "Disciplining the Transnational Mobility of People, Basingstoke: Palgrave Macmillan, pp. 15–40.

Gibbons, Michael; Limoges, Camille; Nowotny, Helga; Schwartzman, Simon; Scott, Peter; Trow, Martin (1994): "The New Production of Knowledge. The dynamics of science and research in contemporary societies", London: Sage.

Griese, Hartmut (Ed.) (1984a): "Der Gläserne Fremde. Bilanz und Kritik der Gastarbeiterforschung und der Ausländerpädagogik", Opladen: Leske+Budrich.

Griese, Hartmut (1984b): "Vorwort und Einleitung", in: Hartmut Griese (Ed.), "Der Gläserne Fremde. Bilanz und Kritik der Gastarbeiterforschung und der Ausländerpädagogik", Opladen: Leske+Budrich.

Groß, Thomas (2010): "Ressortforschung, Agenturen und Beiräte – Zur notwendigen Pluralität der staatlichen Wissensinfrastruktur", in: Hans-Christian Röhl (Ed.), "Wissen – zur Kognitiven Dimension des Rechts", pp. 135–155.

Gusy, Christoph; Müller, Sebastian (2012): "Social Construction of Heterogeneity Indicators and their Relationship to Law. The Example of Guiding Principles in Immigration Law", in: *SFB Working Paper Series* (9).

Gütlhuber, Thomas; Schimany, Peter (2013): "Die Forschungsgruppe im Bundesamt für Migration und Flüchtlinge", in: Bundesamt für Migration und Flüchtlinge (Ed.), "60 Jahre Bundesamt für Migration und Flüchtlinge: Festschrift Forschung", Nürnberg, pp. 19–33.

Haller, Michael (2017): "Die 'Flüchtlingskrise' in den Medien. Tagesaktueller Journalismus zwischen Meinung und Information", in: *OBS Arbeitsheft* (93).

Haraway, Donna (1988): "Situated Knowledges: The Science Question in Feminism and the Privilege of Partial Perspective", in: *Feminist Studies* 14, pp. 575–599.

Harris, Roger (2015): "The Impact of Research on Development Policy and Practice: This Much We Know", in: Arul Chib, Julian May, Roxana Barrantes (Eds.), "Impact of Information Society Research in the Global South", Singapore: Springer, pp. 25–45.

Heckmann, Friedrich (2012): "Willkommenskultur. Was ist das, und wie kann sie entstehen und entwickelt werden?" (= *efms paper* 7), Bamberg.

Heckmann, Friedrich (2013): "Zur Entstehung und Bedeutung der Migrations- und Integrationsforschung in Deutschland", in: Bundesamt für Migration und Flüchtlinge (Ed.), "60 Jahre Bundesamt für Migration und Flüchtlinge: Festschrift Forschung", Nürnberg, pp. 33–43.

Heckmann, Friedrich; Wiest, Delia (2015): "Research-Policy Dialogues in Germany", in: Peter Scholten, Han Entzinger, Rinus Penninx, Stijn Verbeek (Eds.), "Integrating Immigrants in Europe", Heidelberg: Springer, 185-211.

Herbert, Ulrich (2000): "Geschichte der Ausländerpolitik in Deutschland: Saisonarbeiter, Zwangsarbeiter, Gastarbeiter, Flüchtlinge," bpb: Berlin.

Hernández Aguilar, Luis Manuel (2014): "Rassifizierte Subjekte und die Sprache der Toleranz. Die Deutsche Islam Konferenz und die Institutionalisierung des Islam in Deutschland", In Gudrun Hentges, Kristina Nottblohm, Mechtild Jansen, Jamila Adamou (Eds.), "Sprache – Macht – Rassismus", Berlin: Metropol, pp. 265–285.

Hernández Aguilar, Luis Manuel (2018): "Governing Muslims and Islam in Contemporary Germany. Race, Time, and the German Islam Conference", Leiden: Brill.

Hess, Sabine (2014): "Das Regieren der Migration als wissensbasierte Netzwerkpolitik", In Jens Adam, Asta Vonderau (Eds.), "Formationen des Politischen", Bielefeld: transcript, 241-273.

Hetfleisch, Gerhard (2013): "Migrationsforschung als Apologie herrschender Verhältnisse am Beispiel Hartmut Essers", in: Paul Mecheril, Oscar Thomas-Olalde, Claus Melter, Susanne Arens, Elisabeth Romaner (Eds.), "Migrationsforschung als Kritik? Konturen einer Forschungsperspektive", Wiesbaden: Springer, pp. 213–231.

Hetfleisch, Gerhard (2017): "Von Migrationsforschung und Legitimationswissenschaft", in: Lena Karasz (Ed.), "Migration und die Macht der Forschung. Kritische Wissenschaft in der Migrationsgesellschaft", Wien: OGB Verlag, pp. 91–103.

Hollifield, James (2004): "The Emerging Migration State", in: *International Migration Review* 38, pp. 885–912.

Hönekopp, Elmar (Ed.) (1987a): "Aspekte der Ausländerbeschäftigung in Deutschland. Institut für Arbeitsmarkt- und Berufsforschung, Nürnberg (Beiträge zur Arbeitsmarkt- und Berufsforschung, 114).

Hönekopp, Elmar (1987b): "Rückkehrförderung und Rückkehr ausländischer Arbeitnehmer. Ergebnisse des Rückkehrförderungsgesetzes, der Rückkehrhilfe-Statistik und der IAB-Rückkehrerbefragung", in: Elmar Hönekopp (Ed.), "Aspekte der Ausländerbeschäftigung in Deutschland" (Beiträge zur Arbeitsmarkt- und Berufsforschung, 114), pp. 287–341.

Jenkins, Richard (2000): "Categorization: Identity, Social Process and Epistemology", In *Current Sociology* 48, pp. 7–25.

Joppke, Christian (2001): "The Evolution of Alien Rights in the United States, Germany, and the European Union", in: Thomas Alexander Aleinikoff (Ed.), "Citizenship Today: Global perspectives and practices", Washington: Brookings Institution Press, pp. 36–62.

Jureit, Ulrike (2012): "Das Ordnen von Räumen. Territorium und Lebensraum im 19. und 20. Jahrhundert", Hamburg: Hamburger Edition.

Karabell, Zachary (2014): "The Leading Indicators. A Short History of the Numbers that Rule our World", New York: Simon & Schuster.

Kelek, Necla (2006): "Teilnahme von muslimischen Kindern, insbesondere Mädchen, am Sport-, Schwimm- und Sexualkundeunterricht an staatlichen Schulen, Teilnahme an Klassenfahrten", Bundesamt für Migration und Flüchtlinge, Berlin.

Kerpal, Marlene (Ed.) (2003): „50 Jahre – Behörde im Wandel. 1953-2003," Nürnberg.

Kleist, J. Olaf (2018): "Flucht: Forschung und Transfer. Institut für Migrationsforschung und Interkulturelle Studien", Osnabrück.

Klekowski von Kloppenfels, Amanda (2003): "Willkommene Deutsche oder tolerierte Fremde? Aussiedlerpolitik und -verwaltung in der Bundesrepublik Deutschland seit den 1950er Jahren", in: Jochen Oltmer (Ed.), "Migration steuern und verwalten. Deutschland vom späten 19. Jahrhundert bis zur Gegenwart", Osnabrück: V&R unipress, pp. 399–422.

Kraler, Albert; Perchinig, Bernhard (2017): "Der Nexus von Politik und Forschung im Bereich Migration, Integration und Asyl", in: Lena Karasz (Ed.), "Migration und die Macht der Forschung. Kritische Wissenschaft in der Migrationsgesellschaft", Wien: OGB Verlag, pp. 63–90.

Kraler, Albert; Reichel, David; Entzinger, Han (2015): "Integrating Immigrants in Europe. Research-Policy Dialogues", in: Peter Scholten, Han Entzinger, Rinus Penninx, Stijn Verbeek (Eds.), "Integrating Immigrants in Europe", Heidelberg: Springer, pp. 39–58.

Kratzer, Vinzenz (2018a): "History of State Migration Research in Germany", In: *AEMI Journal* (15-16), S. 89–103.

Kratzer, Vinzenz (2018b): "Konzept und Entwicklung der europäischen Transitlandpolitik", in: Ursula Bitzegeio, Frank Decker, Sandra Fischer, Thorsten Stolzenberg (Eds.), "Flucht, Transit, Asyl. Interdisziplinäre Perspektiven auf ein Europäisches Versprechen", Bonn: Dietz-Verlag, pp. 133–149.

Kreienbrink, Axel (2013): „60 Jahre Bundesamt für Migration und Flüchtlinge im Kontext der deutschen Migrationspolitik", in: *ZAR Zeitschrift für Ausländerrecht und Ausländerpolitik* 33, pp. 397–448.

Kreienbrink, Axel; Worbs, Susanne (2015): "Zehn Jahre Migrations- und Integrationsforschung im Bundesamt für Migration und Flüchtlinge", in: *ZAR Zeitschrift für Ausländerrecht und Ausländerpolitik* 35, pp. 325–335.

Kreienbrink, Axel; Worbs, Susanne (2018): "'Amtliche' Wissenschaft im Schnittfeld verschiedener Öffentlichkeiten. Das Forschungszentrum des Bundesamtes für Migration und Flüchtlinge", in: Stefan Sele, Annette

Treibel (Eds.), "Öffentliche Gesellschaftswissenschaften: Grundlagen, Anwendungsfelder und neue Perspektiven", Wiesbaden: Springer, 237-254.

Kron, Thomas; Winter, Lars (2011): "Die radikale Unbestimmtheit des Sozialen", in: Daniel Fischer, Wolfgang Bonß, Thomas Augustin, Felix Bader, Michaela Pichlbauer, Dominikus Vogl (Eds.), "Uneindeutigkeit als Herausforderung – Risikokalkulation, Amtliche Statistik und die Modellierung des Sozialen", Neubiberg, pp. 187–215.

Kühn, Heinz (1979): "Stand und Weiterentwicklung der Integration der ausländischen Arbeitnehmer und ihrer Familien in der Bundesrepublik Deutschland. Memorandum des Beauftragten der Bundesregierung", Bonn.

Lanz, Stephan (2007): "Berlin aufgemischt. abendländisch-multikulturell-kosmopolitisch? Die politische Konstruktion einer Einwanderungsstadt", Bielefeld: transcript.

Latour, Bruno; Woolgar, Steve (1986): "Laboratory Life. The Construction of Scientific Facts", Princeton: Princeton University Press.

Lederer, Harald (2004): "Indikatoren der Migration. zur Messung des Umfangs und der Arten von Migration in Deutschland unter besonderer Berücksichtigung des Ehegatten- und Familiennachzugs sowie der illegalen Migration", Bamberg.

Leibnitz-Institut für Sozialwissenschaften (2010): "Migration und ethnische Minderheiten", (= Sozialwissenschaftlicher Fachinformationsdienst, 2010/2), Bonn.

Lemberg, Eugen; Edding, Friedrich (Eds.) (1959): "Die Vertriebenen in Westdeutschland. Ihre Eingliederung und ihr Einfluss auf Gesellschaft, Wirtschaft, Politik und Geistesleben", Kiel: Ferdinand Hirt.

Lindblom, Charles (1959): "The Science of 'Muddling Through'", in: *Public Administration Review* 19, pp. 79–88.

Lundgreen, Peter (1986): "Staatliche Forschung in Deutschland: 1870-1980", Frankfurt: campus.

Magistrat der Landeshauptstadt Wiesbaden (2012): "Integrationsbericht 2012", Wiesbaden.

Mangset, Marte; Asdal, Kristin (2018): "Bureaucratic power in note-writing. Authoritative expertise within the state", in: *The British journal of sociology* 70, 1-20.

Marfaing, Laurence (2011): "Wechselwirkungen zwischen der Migrationspolitik der Europäischen Union und Migrationsstrategien in Westafrika", in: Bundesamt für Migration und Flüchtlinge (Ed.), "Potenziale der Migration zwischen Afrika und Deutschland" (= Beiträge zu Migration und Integration 2), pp. 63–89.

MARPLAN (1995): "Ausländer in Deutschland: 1970 bis 1995. Ausgewählte Ergebnisse der MARPLAN-Ausländerforschung", Offenbach.

Massey, Douglas; Arango, Joaquin; Hugo, Graeme; Kouaouci, Ali; Pellegrino, Adela; Taylor, J. Edward (1998): "Worlds in Motion. Understanding International Migration at the End of the Millennium", Oxford: Clarendon Press.

Mayr, Katharina; Nassehi, Armin; Hangen-Demszky, Alma von der (2011): "Knowledge as practice", Available online at www.knowandpol.eu.

Mecheril, Paul; Kalpaka, Annita; Melter, Claus; Dirim, İnci; Castro Varela, María do Mar (Eds.) (2010): "Migrationspädagogik", Weinheim, Basel: Beltz Verlag.

Mecheril, Paul; Thomas-Olalde, Oscar; Melter, Claus; Arens, Susanne; Romaner, Elisabeth (2013): "Migrationsforschung als Kritik? Erkundung eines epistemischen Anliegens in 57 Schritten", in: Paul Mecheril, Oscar Thomas-Olalde, Claus Melter, Susanne Arens, Elisabeth Romaner (Eds.), "Migrationsforschung als Kritik? Konturen einer Forschungsperspektive", Wiesbaden: Springer, pp. 7–55.

Mehrländer, Ursula (1987): "Ausländerforschung 1965 bis 1980. Fragestellungen, theoretische Ansätze, empirische Ergebnisse" (=Ausländerforschung und Ausländerpolitik, 4), Bonn: Neue Gesellschaft GmbH

Messerschmidt, Reinhard (2017): "Demografisierung des Gesellschaftlichen. Eine diskursanalytische Bestandsaufnahme", in: Julian Hamann, Vincent Gengnagel, Jens Maeße, Alexander Hirschfeld (Eds.), "Macht in Wissenschaft und Gesellschaft. Diskurs- und feldanalytische Perspektiven", Wiesbaden: Springer, pp. 319–358.

Middelmann, Werner (1959): "Entstehung und Aufgaben der Flüchtlingsverwaltung", in: Eugen Lemberg, Friedrich Edding (Eds.), "Die Vertriebenen in Westdeutschland. Ihre Eingliederung und ihr Einfluss auf Gesellschaft, Wirtschaft, Politik und Geistesleben", Kiel: Ferdinand Hirt.

Müller, Tobias (2018): "Sicherheitswissen und Extremismus. Definitionsdynamiken in der deutschen Islampolitik", in: Schirin Amir-Moazami (Ed.), "Der inspizierte Muslim. Zur Politisierung der Islamforschung in Europa", Bielefeld: transcript, pp. 185–214.

Nahm, Peter Paul (1959): "Der Wille zur Eingliederung", in: Eugen Lemberg, Friedrich Edding (Eds.), "Die Vertriebenen in Westdeutschland. Ihre Eingliederung und ihr Einfluss auf Gesellschaft, Wirtschaft, Politik und Geistesleben", Kiel: Ferdinand Hirt.

Nassehi, Armin; Hangen-Demszky, Alma von der; Mayr, Katharina (2009): "The Amendement of the Bavarian Education Law in 2003. A Long Way towards Inclusion", in: *KNOWandPOL Working Paper* (10).

Nellner, Werner (1959): "Grundlagen und Hauptergebnisse der Statistik", in: Eugen Lemberg, Friedrich Edding (Eds.), "Die Vertriebenen in Westdeutschland. Ihre Eingliederung und ihr Einfluss auf Gesellschaft, Wirtschaft, Politik und Geistesleben", Kiel: Ferdinand Hirt.

Neumann, Iver (2012): "At Home with the Diplomats. Inside a European Foreign Ministery", Ithaca, London: Cornell University Press.

Nowotny, Helga; Scott, Peter; Gibbons, Michael (2003): „'Mode 2' Revisited: The New Production of Knowledge", in: *Minerva* 41, pp. 179–194.

Ozga, Jenny; Grek, Sotiria; Lawn, Martin (2009): "The New Production of Governing Knowledge: Education Research in England", in: *Soziale Welt* 60, pp. 353–369.

Parisius, Bernhard (2003): "Flüchtlingsverwaltung in der britischen und amerikanischen Besatzungszone", in: Jochen Oltmer (Ed.), "Migration steuern und verwalten. Deutschland vom späten 19. Jahrhundert bis zur Gegenwart", Osnabrück: V&R unipress, pp. 253–268.

Pastore, Ferruccio (2011): "The Migration Impact of North African Revolutions and What it Reveals about Europe", in: *EuroMesco Brief* (6).

Piore, Michael (1979): "Birds of Passage", Cambridge: Cambridge University Press.

Rose, Nikolas (1991): "Governing by Numbers. Figuring out Democracy", in: *Accounting, Organizations, and Society* 16 (7), pp. 673–692.

Rosenberg, Daniel (2013): "Data before the Fact", in: Lisa Gitelman (Ed.), "Raw Data" is an oxymoron", Cambridge, MA: The MIT Press, pp. 15–40.

Sachverständigenrat deutscher Stiftungen für Integration und Migration (2017): "Die Messung von Integration in Deutschland und Europa. Möglichkeiten und Grenzen bestehender Integrationsmonitorings", Berlin.

Sachverständigenrat für Zuwanderung und Integration (2004): "Migration und Integration – Erfahrungen nutzen, Neues wagen", Berlin.

Salentin, Kurt (2014): "Sampling the Ethnic Minority Population in Germany. The Background to 'Migration Background'", in: *methods, data, analyses* 8, pp. 25–52.

Sanderson, Ian (2002): "Making Sense of 'What Works': Evidence Based policy-making as Instrumental Rationality?", In *Public Policy and Administration* 17, pp. 61–75.

Schepelern Johansen, Brigitte; Spielhaus, Riem (2018): "Die Vermessung der Muslime. Ein Jahrzehnt quantitativer Forschung zu Muslimen in Westeuropa", in: Schirin Amir-Moazami (Ed.), "Der inspizierte Muslim. Zur Politisierung der Islamforschung in Europa", Bielefeld: transcript, 125-157.

Schiffauer, Werner (2018): "Ethos und Wissensproduktion bei Sicherheitsbürokratien", (= *Anton Wilhelm Amo Lectures,* 5), Halle: Martin-Luther-Universität Halle-Wittenberg.

Schimany, Peter; Schick, Herrmann (2009): "Migrationsforschung im Spiegel von Datenbanken: Rückblick auf zehn Jahre sozialwissenschaftlicher Fachinformationsdienst 'Migration und ethnische Minderheiten'" in: *So-Fid Informationsdienst Migration und Ethnische Minderheiten, pp.* 11–47.

Schimany, Peter; Schock, Hermann (2012): "Migrations- und Integrationsforschung im Spiegel der Datenbanken 'Sozialwissenschaftliches Forschungsinformationssystem' (SOFIS und 'Sozialwissenschaftliches Literaturinformationssystem' (SOLIS)", in IMIS-Beiträge 41, Osnabrück.

Schmid, Susanne (2011): "Qualitative Einschätzung potenzieller Migration aus Afrika nach Europa aufgrund interkontinentaler Entwicklungsdifferenzen", in: Bundesamt für Migration und Flüchtlinge (Ed.), "Potenziale der Migration zwischen Afrika und Deutschland" (= Beiträge zu Migration und Integration, 2), Nürnberg.

Schneider, Jan (2010): "Modernes Regieren und Konsens. Kommissionen und Beratungsregime in der deutschen Migrationspolitik", Wiesbaden: VS Verlag für Sozialwissenschaften.

Scholten, Peter (2011a): "Constructing Dutch Immigrant Policy. Research-Policy Relations and Immigrant Integration Policy-Making in the Netherlands", in: *The British Journal of Politics & International Relations* 13, pp. 75–92.

Scholten, Peter (2011b): "Framing Immigrant Integration. Dutch Research-Policy Dialogues in Comparative Perspective", Amsterdam: Amsterdam University Press.

Scholten, Peter; Entzinger, Han; Penninx, Rinus (2015a): "Research-Policy Dialogues on Migrant Integration in Europe: Comparison and Conclusions", in: Peter Scholten, Han Entzinger, Rinus Penninx, Stijn Verbeek (Eds.), "Integrating Immigrants in Europe". Heidelberg: Springer, 315-336.

Scholten, Peter; Entzinger, Han; Penninx, Rinus; Verbeek, Stijn (Eds.) (2015b): "Integrating Immigrants in Europe", Heidelberg: Springer.

Schönwälder, Karen (1999): "Ist nur Liberalisierung Fortschritt?". zur Entstehung des ersten Ausländergesetzes der Bundesrepublik", in: Jan Motte, Rainer Ohlinger, Anne von Oswald (Eds.), „50 Jahre Bundesrepublik, 50 Jahre Einwanderung. Nachkriegsgeschichte als Migrationsgeschichte", Frankfurt, New York: campus, pp. 127–144.

Schönwälder, Karen (2003): "Zukunftsblindheit oder Steuerungsversagen? Zur Ausländerpolitik der Bundesregierungen der 1960er und frühen 1970er Jahre", in: Jochen Oltmer (Ed.), "Migration steuern und verwalten. Deutschland vom späten 19. Jahrhundert bis zur Gegenwart", Osnabrück: V&R unipress, pp. 123–144.

Scott, James C. (1998): "Seeing like a State. How Certain Schemes to Improve the Human Condition have Failed", New Haven and London: Yale University Press.

Spielhaus, Riem (2013): "Muslime in der Statistik. Wer ist Muslim und wenn ja wie viele?" Berlin: Mediendienst Integration.

Statistisches Bundesamt (2006): "Bevölkerung Deutschlands bis 2050. 11. koordinierte Bevölkerungsvorausberechnung", Wiesbaden.

Statistisches Bundesamt (2007): "Bevölkerung und Erwerbstätigkeit. Bevölkerung mit Migrationshintergrund. Ergebnisse des Mikrozensus 2005", in: *Fachserie 1 Reihe 2.2.*

Statistisches Bundesamt (2012): "Bevölkerung und Erwerbstätigkeit. Ausländische Bevölkerung. Ergebnisse des Ausländerzentralregisters", in: *DEStatis Fachserie 1 Reihe 2.*

Statistisches Bundesamt (2017): "Bevölkerung und Erwerbstätigkeit 2015. Wanderungen", Wiesbaden.

Straubhaar, Thomas (2003): "Ökonomische Aspekte der Zuwanderung nach Deutschland. In Marlene Kerpal (Ed.), „50 Jahre – Behörde im Wandel. 1953-2003", Nürnberg, pp. 121–137.

Tetlock, Philip (2005): "Expert Political Judgement. How good is it? How can we know?", Princeton, Oxford: Princeton University Press.

Tezcan, Levent (2011): "Der säkulare Muslim. Zur Generierung einer Kategorie im Kontext der Deutschen Islam Konferenz", in: Marianne Krüger-Potratz, Werner Schiffauer (Eds.), "Migrationsreport 2010. Fakten – Analysen – Perspektiven", Frankfurt, New York: campus, pp. 83–108.

Thränhardt, Dietrich (1984): "'Ausländer' als Objekte deutscher Interessen und Ideologien", in: Hartmut Griese (Ed.), "Der Gläserne Fremde. Bilanz und Kritik der Gastarbeiterforschung und der Ausländerpädagogik", Opladen: Leske+Budrich, pp. 103–134.

Unabhängige Kommission "Zuwanderung" (2001): "Zuwanderung gestalten – Integration fördern", Berlin.

Vogel, Dita; Aßner, Manuel (2012): "Expertise 'Umfang, Entwicklung und Struktur der irregulären Bevölkerung in Deutschland'", Bundesamt für Migration und Flüchtlinge: Nürnberg.

Walters, William (2015): "Reflections on Migration and Governmentality", in: *movements. Journal für kritische Migrations- und Grenzregimeforschung* 1.

Walzer, Michael (1987): "Interpretation and Social Criticism" Cambridge/Mass: Harvard University Press.

Weber, Max (1994): "Political Writings", With assistance of Peter Lassman, Ronald Speirs. Cambridge.

Weber, Max (2005): "Bürokratismus", in: Max Weber: "Wirtschaft und Gesellschaft. Die Wirtschaft und die gesellschaftlichen Ordnungen der Mächte. Nachlass", Edited by Edith Hanke. Tübingen: C. J. B. Mohr, pp. 150–234.

Wilpert, Czarina (1984): "International Migration and Ethnic Minorities: New fields for Post-War Sociology in the Federal Republic of Germany", in: *Current Sociology* 33 (3), 305-352.

Wimmer, Andreas (2009): "Hartmut Essers Assimilationsmodell zwischen empirischer Sozialforschung und makrosozialer Theorie", in: Paul Hill, Johannes Kopp, Clemens Kroneberg, Rainer Schnell (Eds.), "Hartmut Essers Erklärende Soziologie. Kontroversen und Perspektiven", Frankfurt: campus, pp. 319–348.

Wimmer, Andreas; Glick Schiller, Nina (2003): "Methodological Nationalism, the Social Sciences, and the Study of Migration: An Essay in Historical Epistemology", in: *International Migration Review* 37, pp. 576–610.

Wollenschläger, Michael (2003): "Deutsche Ausländer- und Asylpolitik in Bewegung." In: Marlene Kerpal (Ed.), 50 Jahre – Behörde im Wandel. 1953 – 2003", Nürnberg: Bundesamt für die Anerkennung Ausländischer Flüchtlinge (11), pp. 40–56.

Worbs, Susanne (2010): "Integration in klaren Zahlen? Ansätze des Integrationsmonitorings in Deutschland", in: *focus Migration Kurzdossier* (16).

Zur Nieden, Birgit (2009):"'...und Deutsch ist wichtig für die Sicherheit!'. Eine kleine Genealogie des Spracherwerbs Deutsch in der BRD", in: Sabine Hess, Jana Binder, Johannes Moser (Eds.), No integration?! Bielefeld: transcript, pp. 123–136.

## List of Interviews

Recorded, Transcripted Interviews:
3 Interviews at the Research Group, Nuremberg, September 2015
1 Interview with a former member of the Research Group, December 2016
1 Interview with a former member of the Research Group's Advisory Board, December 2017
Quoted passages are referenced and translated in the text; the original German text is quoted in the footnotes. All interviews have been anonymized.
Background talks, not taped-recorded and transcripted, documented in the research notes:
1 Interview with a former BAMF official, October 2013
1 Interview with a former Research Group Member, July 2016
1 Interview with a former research associate, February 2014
Quoted passages are referenced to as "Field notes" with the respective dates.

# Social Sciences

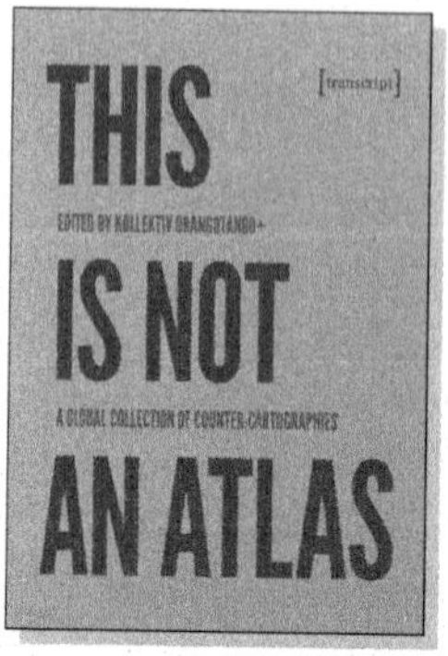


kollektiv orangotango+ (ed.)
**This Is Not an Atlas**
**A Global Collection of Counter-Cartographies**

2018, 352 p., hardcover, col. ill.
34,99 € (DE), 978-3-8376-4519-4
E-Book: free available, ISBN 978-3-8394-4519-8

Gabriele Dietze, Julia Roth (eds.)
**Right-Wing Populism and Gender**
**European Perspectives and Beyond**

April 2020, 286 p., pb., ill.
35,00 € (DE), 978-3-8376-4980-2
E-Book: 34,99 € (DE), ISBN 978-3-8394-4980-6

Mozilla Foundation
**Internet Health Report 2019**
2019, 118 p., pb., ill.
19,99 € (DE), 978-3-8376-4946-8
E-Book: free available, ISBN 978-3-8394-4946-2

**All print, e-book and open access versions of the titles in our list are available in our online shop www.transcript-publishing.com**

# Social Sciences

James Martin
**Psychopolitics of Speech**
**Uncivil Discourse and the Excess of Desire**

2019, 186 p., hardcover
79,99 € (DE), 978-3-8376-3919-3
E-Book:
PDF: 79,99 € (DE), ISBN 978-3-8394-3919-7

Michael Bray
**Powers of the Mind**
**Mental and Manual Labor**
**in the Contemporary Political Crisis**

2019, 208 p., hardcover
99,99 € (DE), 978-3-8376-4147-9
E-Book:
PDF: 99,99 € (DE), ISBN 978-3-8394-4147-3

Dorothee Brantz, Avi Sharma (eds.)
**Urban Resilience in a Global Context**
**Actors, Narratives, and Temporalities**

October 2020, 224 p., pb.
30,00 € (DE), 978-3-8376-5018-1
E-Book: available as free open access publication
PDF: ISBN 978-3-8394-5018-5

**All print, e-book and open access versions of the titles in our list are available in our online shop www.transcript-publishing.com**